CW01113044

BRAIN FOOD

Dr Karl's

BRAIN FOOD

MACMILLAN

First published 2011 in Macmillan by
Pan Macmillan Australia Pty Limited
1 Market Street, Sydney

Copyright © Karl S. Kruszelnicki Pty Ltd 2011

The moral right of the author has been asserted.

All rights reserved. No part of this book may be reproduced or transmitted by any person or entity (including Google, Amazon or similar organisations), in any form or by any means, electronic or mechanical, including photocopying, recording, scanning or by any information storage and retrieval system, without prior permission in writing from the publisher.

National Library of Australia Cataloguing-in-Publication data:

Kruszelnicki, Karl, 1948–

Brain food / Karl Kruszelnicki.

9781742610399 (hbk.)

Food.
Science – Popular Works.
Science – Miscellanea.

500

Cover, internal design and typeset by Xou Creative, www.xou.com.au

Cover and internal photographs of Dr Karl by Mel Koutchavlis; Other internal photographs by Thinkstock; Photograph of Alice Kruszelnicki in bucket and Karl Jnr with Alice in the bath by Karl Kruszelnicki; Internal electron microscopy photography courtesy the Australian Centre for Microscopy & Microanalysis; Sleek Geeks photo taken by Damon Willis, supplied by Cordell Jigsaw Productions; Internal illustrations by Douglas Holgate.

Printed in China by 1010 Printing International Limited

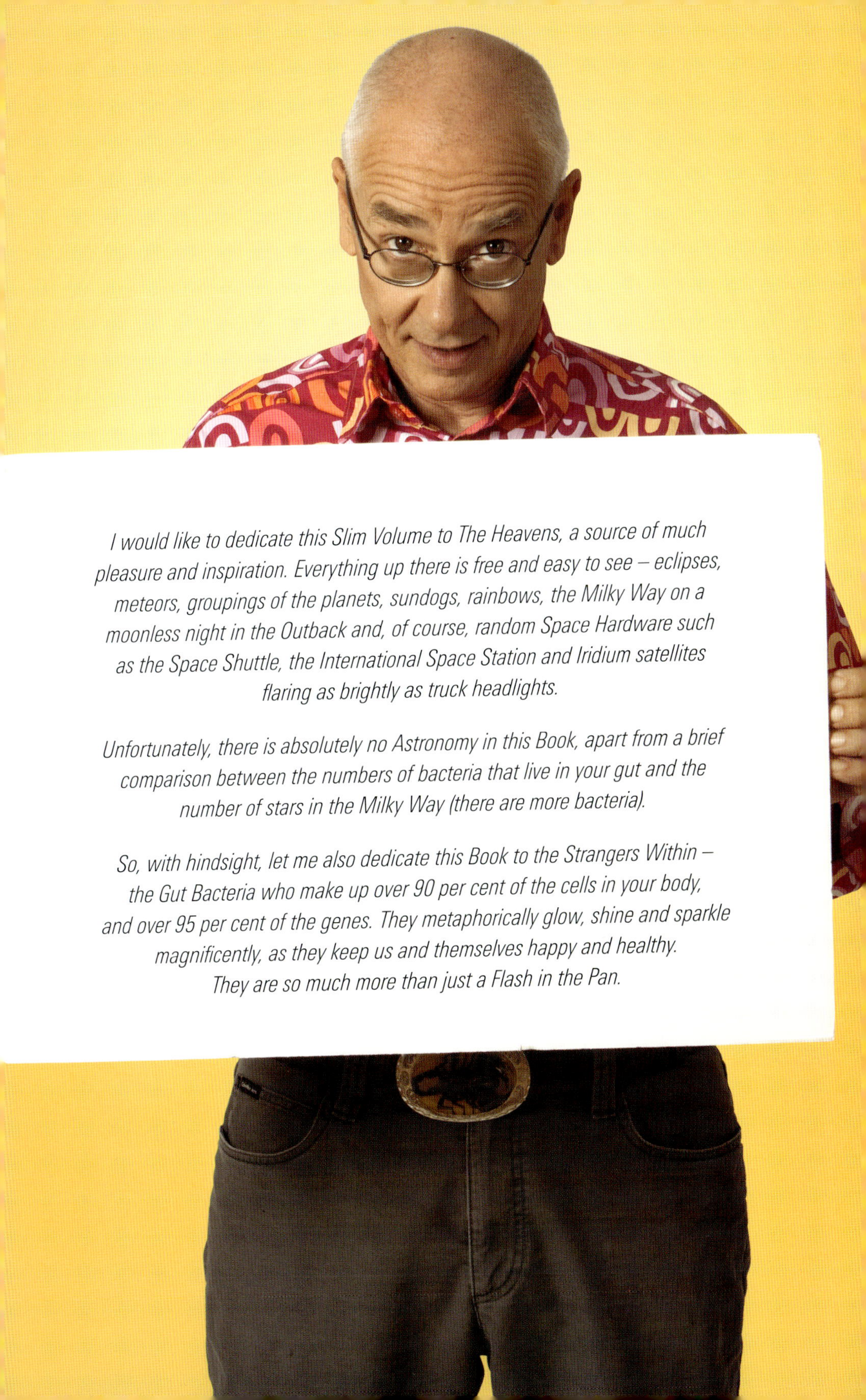
I would like to dedicate this Slim Volume to The Heavens, a source of much pleasure and inspiration. Everything up there is free and easy to see – eclipses, meteors, groupings of the planets, sundogs, rainbows, the Milky Way on a moonless night in the Outback and, of course, random Space Hardware such as the Space Shuttle, the International Space Station and Iridium satellites flaring as brightly as truck headlights.
Unfortunately, there is absolutely no Astronomy in this Book, apart from a brief comparison between the numbers of bacteria that live in your gut and the number of stars in the Milky Way (there are more bacteria).
So, with hindsight, let me also dedicate this Book to the Strangers Within – the Gut Bacteria who make up over 90 per cent of the cells in your body, and over 95 per cent of the genes. They metaphorically glow, shine and sparkle magnificently, as they keep us and themselves happy and healthy. They are so much more than just a Flash in the Pan.

Contents

(WHAT GOES IN, MUST COME OUT)

BRAINY FOUNDATION TICK OF APPROVAL

Up Close and Personal

find the answer on page 229

Moving Foreword

My attitude to food has changed a lot over the years. From childhood, to living in Papua New Guinea as an adult, food was of no consequence. Then, as a hippie, I believed that food was sacred and essential to being well balanced. Much later, as a medical student, I saw it as fuel, and happily lived on soybeans and steamed vegies (for a really big thrill, I'd add caraway seeds).

Today I feel lucky to live with my family who cook meals from scratch nearly every day, and so I enjoy a great diversity of food – and I know that my health benefits from it.

So, with great pleasure and ever-increasing understanding, I took a look at food through the microscope of Science. I was amazed at what I found.

For example, did you know the "Diet Industry" is thriving by selling products that *fail* over 80 per cent of the time? Or that the Big Food Industry is selling "edible" stuff that makes you feel *hungrier* instead of *fuller*? How's that for "un-full-filling"?

Why do some people get ice-cream headaches? What virus can make you fat? Why does the vegetarian Giant Panda have a meat-eater's gut? How can hamburgers kill? And is it true that vodka has no carbs *and* no calories?

My favourite story in this book is about the woman whose life was saved by a transplant of faeces. My cautious publisher couldn't bear it as the book's title, so I snuck it onto the cover, and here as well.

So, turn the pages and learn about the strangers within your gut . . .

DAILY PRESS
FINAL
SPORTS EXTRA
WOMAN SAVED
POO TRANSPL
Atom tests 'affected IQ'
WIFE-ST
Pioneer he
for the unk

The Woman whose Life was Saved by a Poo Transplant

This is the astonishing story of a woman who was dying from a bowel infection, and who was saved by a transplant of faeces – yep, the brown stuff you flush down the toilet.

The 61 year old had been treated with antibiotics for pneumonia. Unfortunately, as sometimes happens, although the antibiotics fixed her pneumonia, they also knocked off many of the good microbes in her gut. This meant that there was an opening for a truly nasty bacterium – *Clostridium difficile* – to move in and take over.

After eight months suffering from the gut infection, the woman was close to death. She had lost 27 kilograms and was confined to a wheelchair because she was so weak. She had to wear a nappy because she had frequent and nasty diarrhoea – every 15 minutes. Nothing her doctors tried worked.

Hippocrates (One of the First Doctors)

Back in 400 BC, Hippocrates knew all about the importance of the bowel. He wrote, "Death sits in the bowels", as well as "bad digestion is the root of all evil".

Well, it sounds about as reasonable as "the love of money is the root of all evil".

Clostridium is a Bad Bacterium. Over the last two decades, the incidence of *Clostridium difficile*–associated disease has increased fifty-fold, and the bacterium has become more virulent. The risk factors for it include increasing age, cancer, use of antibiotics, Inflammatory Bowel Disease (IBD) and being a patient in hospital. The symptoms range from mild diarrhoea up to toxic megacolon, causing death. This Bad Bacterium is also very speedy at developing resistance to antibiotics. Indeed, the disease recurs in some 15–35 per cent of cases. When times are bad, *Clostridium* can turn into extremely hardy spores, which turn back into *Clostridium* bacteria in easier times. These spores are how *Clostridium* is usually transmitted.

So *Clostridium* is becoming more dangerous with each passing year. In the UK in 2005, one in every 250 Death Certificates listed it as a contributing cause of death. In the USA, there are some 400,000 cases of *Clostridium difficile* infection each year, and 6000 deaths, costing the community over $3 billion.

The woman's gastroenterologist, Dr Khoruts from the University of Minnesota, decided on a rather uncommon therapy – faecal transplantation. Yep, it's exactly what you imagine. You get some poo from somebody else, and insert it into the lower part of the gut. This was after he had found that her gut had been "colonised by all sorts of misfits. The normal bacteria just didn't exist in her".

First, Dr Khoruts gave his patient a bowel flush. Then, using a colonoscope, he collected 25 grams of faeces from the gut of his patient's husband of 44 years. The husband was quite healthy. He then mixed the faeces with 300 millilitres of water, and squirted it from a colonoscope into the woman's gut (the right colon).

The next day, amazingly, the woman's diarrhoea stopped. The day after that, she had her first solid bowel motion in some eight months. Day by day, she began to visibly improve, after a previous eight-month-long downward trajectory towards death. After one whole month, the *Clostridium difficile* had left her body and she continued to make a rapid and complete recovery.

This lucky woman's life was saved by a transplant of poo from her husband. It's also called Bacteriotherapy, or Human Probiotic Infusion. You could call it a "transpoosion" – just like a "transfusion", only lumpy . . .

Analysis undertaken two weeks after the transfusion showed that her husband's gut microbes had taken over her gut. In fact, her gut bacteria were now strikingly similar to her husband's.

Dr Janet Jansson, who worked with Dr Khoruts, said that the new microbes, ". . . [were] . . . able to function and cure her disease in a matter of days. I didn't expect it to work. The project blew me away." Somehow, the transfusion of poo from her husband had magically "reset" the microbes in her gut.

As the old saying goes, "If the poo fits, share it . . ."

Unknown Life

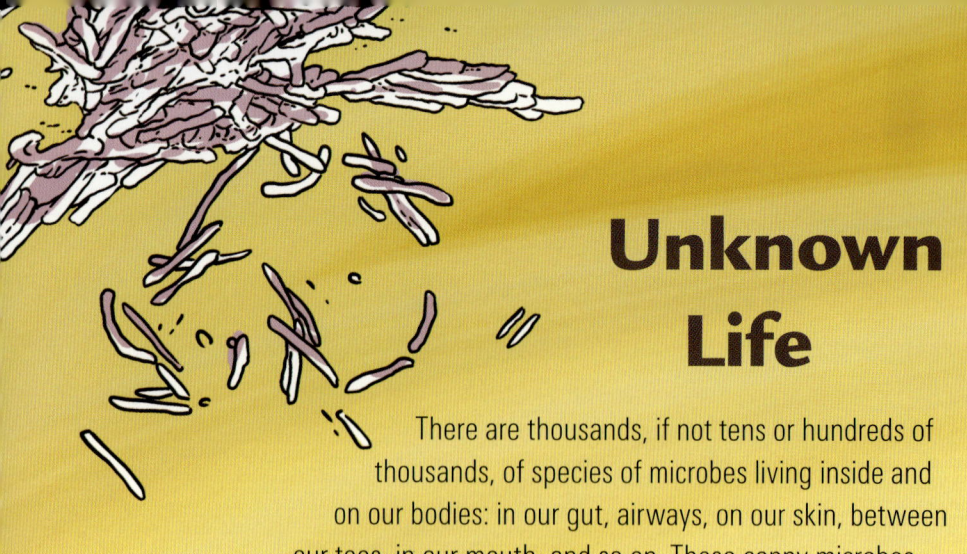

There are thousands, if not tens or hundreds of thousands, of species of microbes living inside and on our bodies: in our gut, airways, on our skin, between our toes, in our mouth, and so on. These canny microbes have colonised every available surface of our body.

But we have no idea what most of them are.

There are at least 500 different known species that live in the mouth alone. The ecosystems are so specific that the microbes that live on the front teeth won't live on the back teeth; and the ones living on the front of a tooth are different from the ones on the back of that same tooth.

There must be something different between our left and right hands, because only 17 per cent of the microbes that live on the left hand also live on the right hand.

Asthmatics have different lung microbes to non-asthmatics. And obese people have different gut microbes to skinny people.

A New Universe Appears

We had no idea of microbes and bacteria until the invention of the microscope in the 1600s. Antonie van Leeuwenhoek, a Dutch lens-grinder, was probably the first to actually see microbes.

Antonie used a blade to scrape scum off his teeth.

A new Universe of Life appeared when his microscope showed tiny swimming creatures.

Nothing is New – 1

We humans have managed livestock for around 10,000 years. And farmers have long seen how cows can get indigestion after a change in diet.

So farmers have treated this by sucking fluid out of the stomach of a healthy cow and feeding it to the sick cow.

This is working at the top end of the gut, not the bottom end, but the principle is the same.

Nothing is New – 2

Back in the 1950s, people began dying of a gut disease called *Pseudomembranous colitis*.

"Colitis" means "colon inflammation" and when doctors looked inside the colon, they saw yellow "pseudomembranes" (fake membranes) on the inner wall of the colon. The symptoms were fever, tummy pain and diarrhoea. In some cases the symptoms got relentlessly worse. In severe cases the death rate was over 75 per cent.

Doctors guessed that the patients' gut bacteria had changed. It was definitely a guess, because back then we didn't have the technology to know.

So, thought the doctors, if (and it was a big IF) the bacteria in the gut had changed, then perhaps they could "reset" them with an infusion of faeces from a healthy person.

The doctors tried it and their guess turned out to be right – many sick patients made miraculous recoveries.

Salmon Colour

In the USA most salmon sold (about 88 per cent) come from salmon farms, not the wild ocean, and this leads to a big difference in the colour of the salmon's flesh.

In the wild, salmon eat little shrimp-like crustaceans called krill. Krill contain a carotenoid (a fat-soluble yellowish dye) which colours the flesh of the salmon that wonderful pink. Salmon are unique in being able to bind this carotenoid to their muscle fibres. Other fish (for example, catfish) as well as baleen whales eat krill, too, but their muscles don't change colour.

A similar carotenoid is released when lobsters and crayfish are cooked, giving their shells that characteristic red.

But getting back to the farmed salmon. They're not fed krill and, although their flesh tastes the same, it's naturally grey. To improve the appearance, the aqua farmers add pink Food Colouring. In the same way you choose paint shades in a paint shop from a colour wheel or colour chart, the farmers pick the final colour they want for their salmon!

The colouring usually comes as the carotenoid astaxanthin, which is either farmed from algae or made synthetically. You can read about it on the packaging label – at least in the USA. Astaxanthin costs about $2000 per kilogram and increases the feed cost of farmed salmon by 10–25 per cent.

The Tube of Life

If you look at another human being running around on the beach, you will see a creature with four limbs, two of which are used for walking, and the other two for balance (and for lifting and carrying stuff).

Human = Doughnut

But if you look at this same person through the eyes of a Topologist (a mathematician who specialises in shapes and geometry), you would see that we humans are basically a fat doughnut. A hollow pipe runs some 10 metres through us, surrounded by everything else. This pipe is the Gut, and it includes the mouth, oesophagus, stomach, small intestine, large intestine and rectum/anus. One end of the pipe is the mouth, the other end is the anus.

The gut is our Tube of Life, because it keeps us alive. It is our machine to process food. (By the way, people have other body holes – a huge number of sweat glands, two nostrils, two ear canals, and openings for the kidney and reproductive tract. But the gut is the only hole that continues right through the body, and comes out on the other side.)

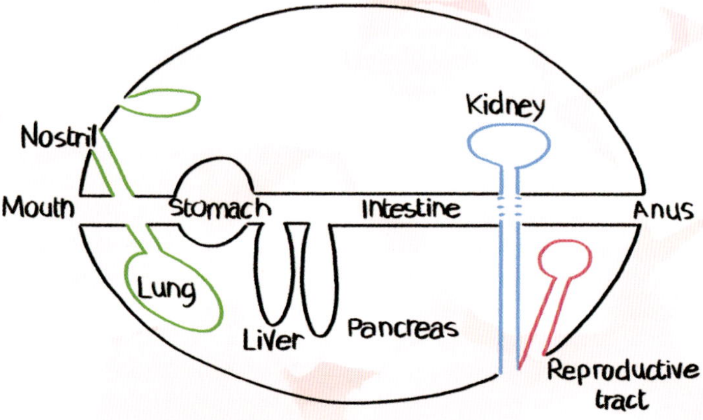

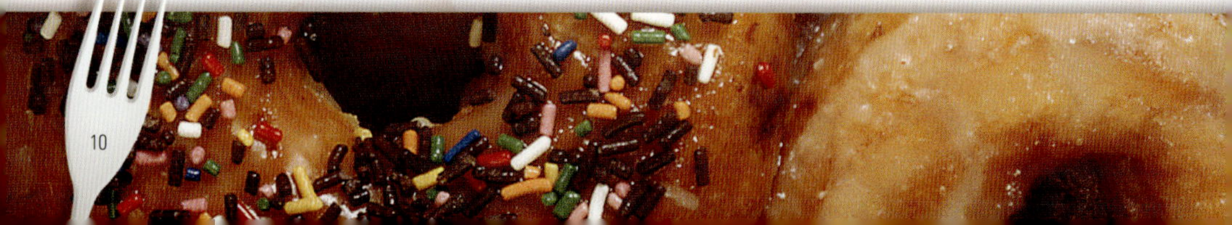

Why We Eat

When we eat food, we want to turn that food into a mix of (1) energy and (2) raw materials or building blocks for the body to use.

There are three main elements in food:

(1) carbohydrates

(2) proteins

(3) fats.

There are also vitamins, minerals and other stuff, but for now, I'll stick to the basic carbs, proteins and fats.

We eat because we are "hungry" (whatever "hungry" is) and because eating is a very pleasant experience.

That is part of the story.

We also eat because there's energy in the chemical bonds that join the atoms of our food together – and we need that energy. Yes, our Tube of Life will break large groups of atoms into successively smaller and smaller groups of atoms. As these chemical bonds are torn apart, our Tube of Life extracts this energy for our body to use.

In addition, we eat because some of these smaller groups of atoms are potential building blocks that our body can use. So our Tube of Life grabs these potential building blocks, and uses them to build stuff – like red blood cells, or bones, or parts of the immune system.

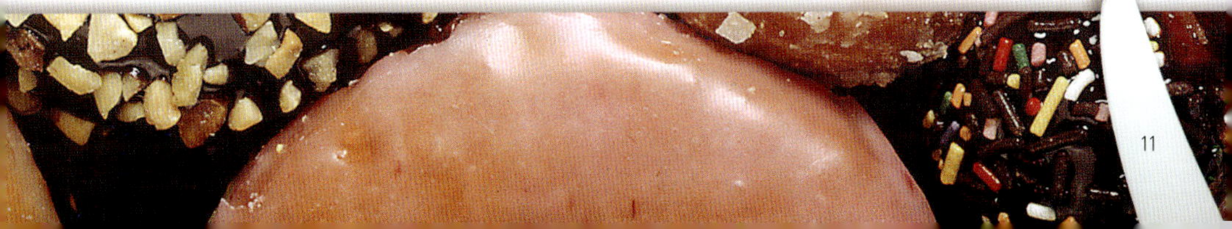

Three Basic Foods

Our first Basic Food is the Carbohydrate (carb for short). Carbs are just a bunch of simple sugars stuck together.

A simple sugar is just a group of little atoms (carbon, hydrogen and oxygen) arranged in a ring. If you stick a bunch of these rings together, you get a carbohydrate.

Our second Basic Food is Protein. You start off with an amino acid. There are about 22 different amino acids that our bodies can make. Surprisingly, amino acid Number 22 (pyrrolysine) was discovered as recently as 2002, and its production process worked out only in early 2011.

Proteins are just a bunch of amino acids joined up. It's fairly arbitrary, but a string of fewer than 50 amino acids is called a "peptide", and if the string is longer than 50, it's generally called a "protein".

Our final Basic Food are the Fats. The fat molecule is a simple one. Typically, there is a glycerol backbone, with three fatty acids hanging off it. It looks a little like the letter "E".

The Elements of Life

There are about 90 different natural elements that make up the universe around us. Our basic foods are made up from just four of these 90 elements.

Both fats and carbohydrates are made from carbon, hydrogen and oxygen. Proteins are also made from carbon, hydrogen and oxygen, but they have an extra element – nitrogen.

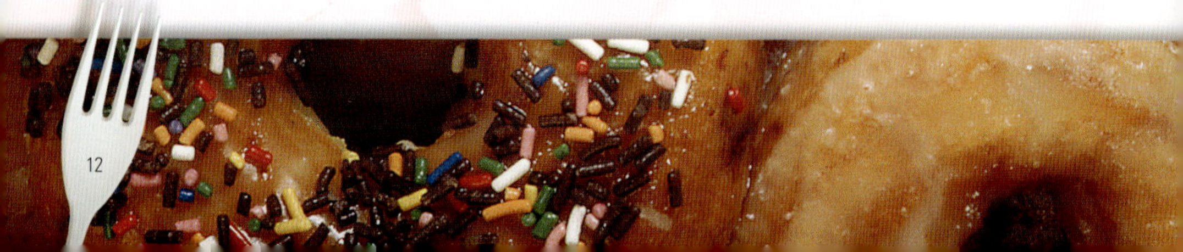

The Mouth Processes Food . . .

The process of breaking down our food starts in the mouth, where we rip, tear and grind our food, using the 32 little white things we call teeth.

The food has to be turned into a mush, so the three salivary glands in your mouth deliver about one litre of saliva each day.

Here's the first weird thing. Saliva contains an enzyme that begins to digest the carbs in your food (not the proteins or fats, just the carbs) immediately. It breaks down enormously long chains of thousands of these sugar rings joined together into shorter chains. These shorter chains have just a couple of these sugar rings stuck together.

So if you've eaten some carbs with a big surface area (say, light, fluffy white bread), the long chains get broken down into these short chains within two minutes. This happens only with carbohydrates – the proteins and the fats are not broken down in the mouth.

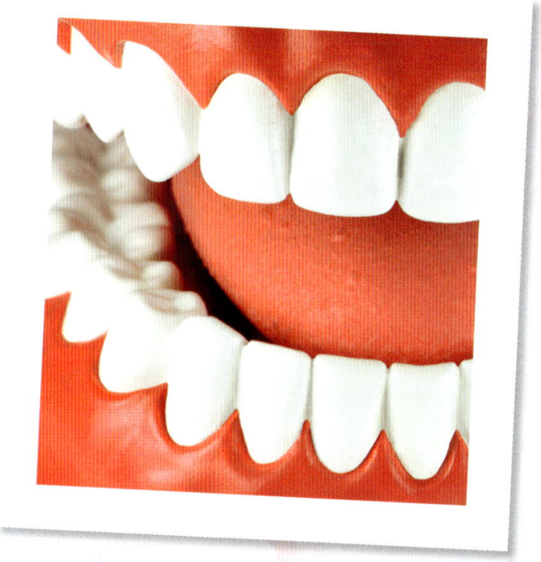

Then you swallow the mush and it goes down the oesophagus into the stomach.

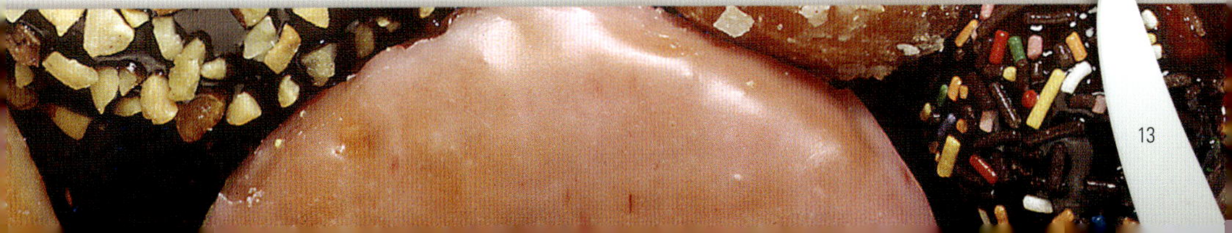

Can you Drink Upside Down?

It's true, you can swallow and drink any liquid while upside down. The muscle wall of the oesophagus has coordinated waves of compression sweeping along its length. These waves will push the contents of the oesophagus towards the stomach.

This coordinated muscular activity is easily strong enough to overcome the force of gravity.

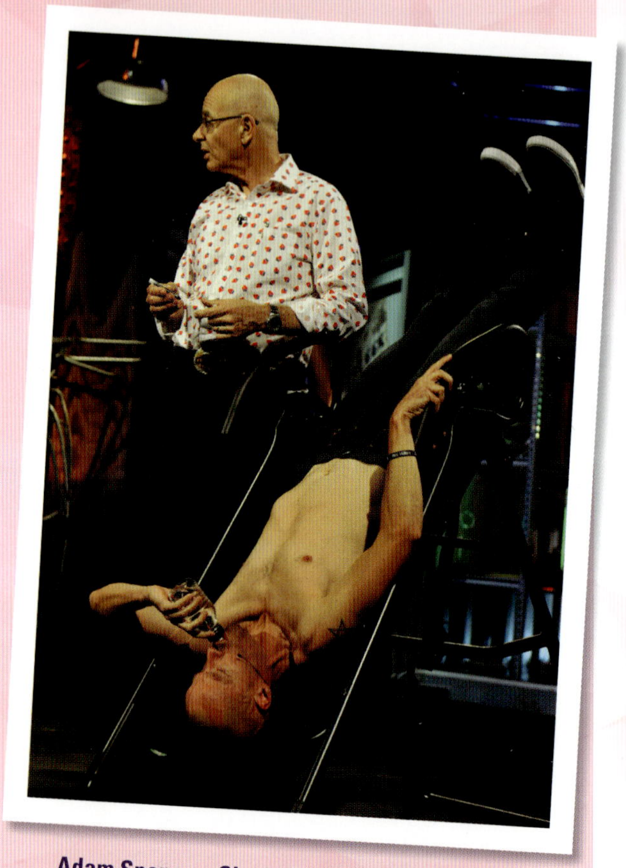

Adam Spencer, *Sleek Geeks 2* (ABC1), drinks water upside down, while Dr Karl explains how.

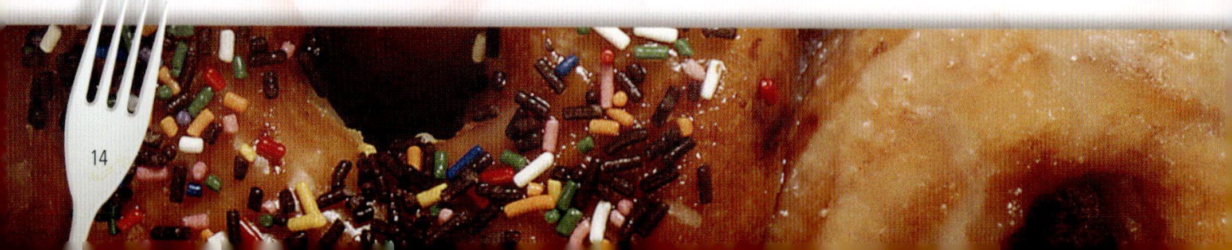

The Stomach Processes Food . . .

The stomach does several things, including grinding and dissolving.

First, it has a powerful muscular wall, which springs into action and starts mushing up and grinding down the food. This increases the surface area of the food even further. This is important, because the chemicals that break down your meal can act only on what they can touch, which is a "surface".

Second, glands in the stomach wall make hydrochloric acid. This acid provides the correct acidity for special enzymes to be released and become active in the stomach. These enzymes, which include "pepsin", then start breaking down proteins.

These enzymes act on any long chains of proteins that they can find, and break them down into lots of smaller chains of amino acids.

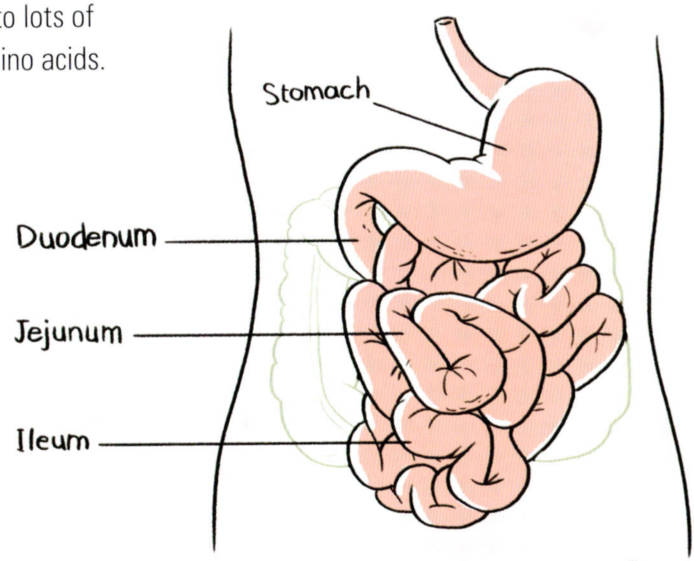

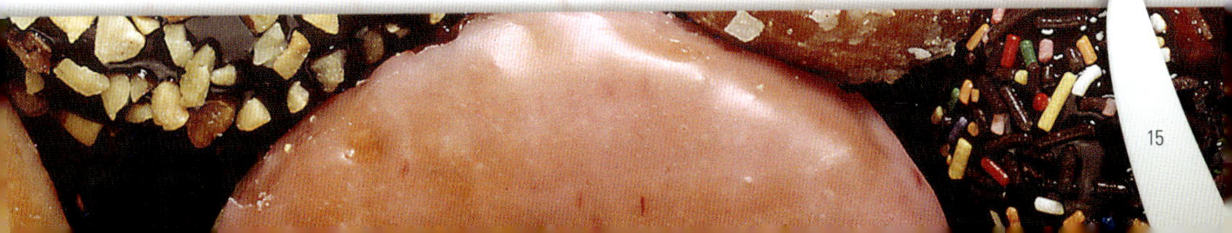

Pepsin in the Stomach

In the stomach, the enzyme called "pepsin" attacks the proteins in meat.

Pepsin has two main pathways of activity.

First, it attacks collagen. Collagen makes up the main bulk of the connective tissue that joins the cells of the meat together. This allows the pepsin to get into the cells of the meat. Second, once pepsin gets inside the cells, it starts breaking down the long chains of amino acids into smaller chains.

Pepsin was the very first animal enzyme to be discovered. In 1836, its discoverer, the German physiologist Theodor Schwann, coined its name "pepsin" from the Greek word "pepsis", which means "digestion". (Reminds me of a commercial soft drink . . .)

Pepsin is one of the three major enzymes that break down proteins in our diet. The other two, trypsin and chymotrypsin, are manufactured in the pancreas and squirted into the duodenum. Each of these three enzymes is specialised for breaking up proteins at different places along the chain of amino acids.

Pepsin works only in acid environments with a pH less than 5.0. Antacids (often taken to relieve tummy pain) can temporarily make the stomach unsuitable for pepsin to work, by inactivating stomach acids.

For a while, pepsin was also blended into chewing gum! Dr Edward E Beeman did this with his Beeman's Gum, with the claim that it was "a preventative or relief for digestion", and that it would "cure Indigestion or Sea-sickness".

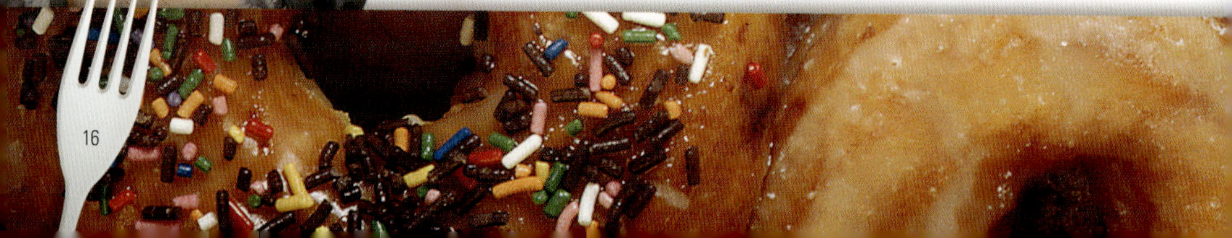

The Small Intestine Processes Food . . .

The next step in the process is when the mush gets squirted out of the stomach and into the small intestine. The small intestine has three parts – the "duodenum", the "jejunum" and the "ileum". The rather short duodenum gets its name because it's as long as 12 fingers put side-by-side (the name comes from the Latin *Duodenum Digitorum* meaning "the width of twelve fingers"). The, jejunum is much longer at about 2.5 metres long, while the ileum is longer again, at about 3.5 metres long.

The jejunum is where the already-broken down chunks of carbs, proteins and fats get broken down into even smaller chunks. The jejunum releases various enzymes (amylase to break down carbohydrates, proteases to break down proteins, and lipases to break down lipids or fats) that work like mini-chemical chainsaws. This is to make the broken-down chunks tiny enough to cross the wall of the gut, and into the pick-up pipes waiting on the other side. These pick-up pipes then feed into blood vessels that carry the components of the sugars, proteins and fats to the liver.

The wall of the gut in the jejunum is covered with many separate layers. Only very small chemicals can get through these many layers.

And by the way, every broken-down bit of carbohydrate, protein and fat has to travel through a cell to get to the other side.

Anything bigger than very small chemicals (viruses and bacteria are too big) can't usually get through this thick wall, which acts as a "filter".

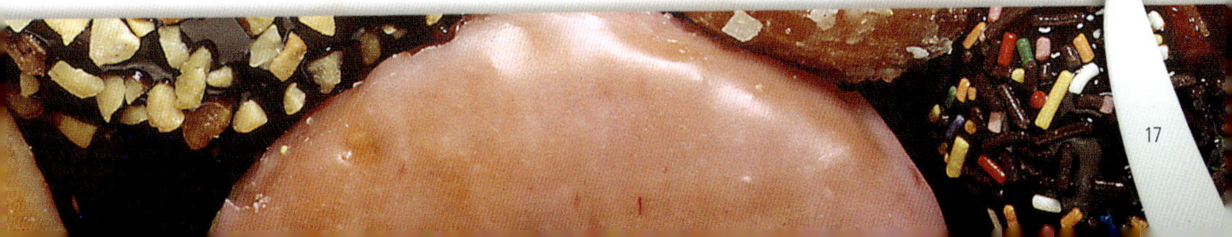

Break-down Foods

Now let's talk about how your food gets absorbed.

The carbohydrate chains get shortened down to just one or two sugar rings, which can travel through the gut wall into the blood vessels on the other side.

The proteins get shortened down to chunks of one or two amino acids, which can also travel through the gut wall into the blood vessels.

The fats are shaped like the letter "E" – three fatty acids hanging off a glycerol backbone. Enzymes chop fats into three smaller chunks: a monoglyceride and two free fatty acids. These are each tiny enough to travel through the wall of the gut. They will first diffuse into a "lacteal", which is a pipe that belongs to the lymphatic system, not the blood vessels. But the fats will eventually end up in the blood vessel system.

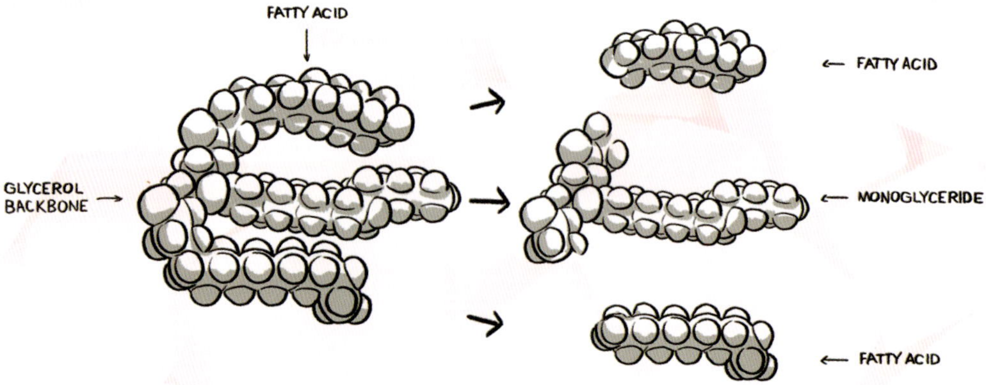

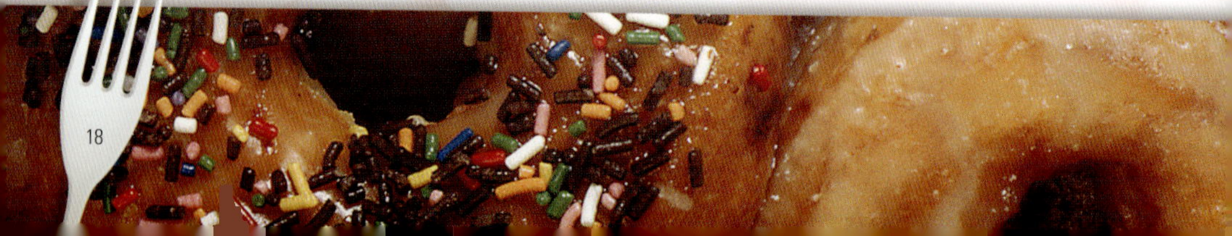

Increase the Surface Area

There are untold trillions and quadrillions of tiny molecules on the inside of the gut. For us to ultimately get energy and building blocks from our food, these molecules all have to "kiss" the inside wall of the Tube of Life, and diffuse through.

You need a lot of surface area to transmit so many molecules. So the small intestine is not just a smooth tube with a constant diameter. If it were, it would have a total surface area of just half a square metre.

The first trick to increase the surface area is that the inner wall is arranged into folds.

These folds or pockets are called valvulae conniventes (or Folds of Kerckring). They protrude as much as 8 millimetres into the internal "lumen", or hollow part of the Tube of Life. These folds increase the surface area by a factor of three.

In turn, the folds are made of many millions of "fingers" or "villi" about 1 millimetre high. Each "finger" or villus has an internal network of tiny blood vessels (and other vessels) to carry away the nutrients. Having the villi or fingers present increases the total available surface area by a factor of ten.

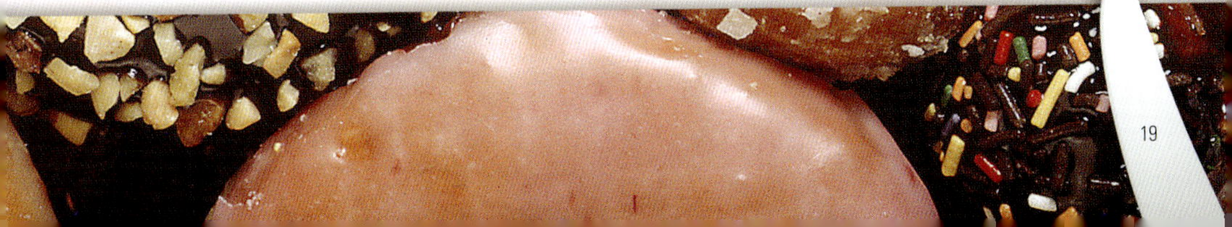

These villi are then covered by even tinier fingers, called "microvilli". There might be 1000 of these even tinier fingers to each cell. The microvilli are about a millionth of a metre long and a tenth of a millionth of a metre wide. They increase the surface area available for absorption by a further factor of 20.

All these folds, fingers and finer fingers increase the surface area of the small intestine from one half of a square metre, to 200 square metres. This is enormous – roughly equal to the size of a tennis court.

The Food Enters the Bloodstream

So now let's talk about how to get the different types of food from inside the tube of the gut into the liver.

The sugars and amino acids leave the small intestine and just diffuse through the gut wall into a blood vessel.

But the fats enter a pipe called a "lacteal", which after a lot of mucking around, delivers them to the liver. These chemicals that you eat – the carbs, proteins and fats – all end up in the liver.

They have to go to the liver first, because if they went straight into the general bloodstream, they would kill you!

These tiny particles (that were once your meal) would deliver a huge "osmotic shock" to the blood after each meal – if you didn't first send them to the liver. Imagine if every time you ate you risked dying. If there were more food particles than normal, water would exit the red blood cells, leaving them all shrivelled and wrinkled, and go into the bloodstream. The red blood cells couldn't survive. But if there were too few food particles, then the water would leave the bloodstream and enter the red blood cells. They would get very plump, and burst open. Either scenario (too many particles immediately after a meal, or too few) can kill you.

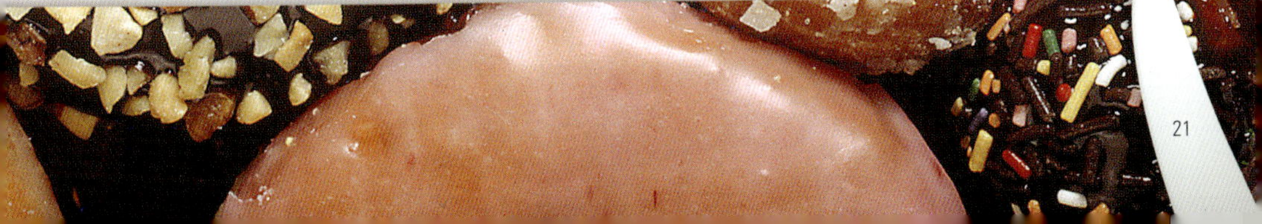

The Lovely Liver

Luckily, all the nutrients from every meal you eat go into the liver before they get into the general circulation.

The liver decides what to do with the carbs, proteins and fats.

- It might break them down for energy.
- It might release some of them into the bloodstream unchanged.
- It might modify some of them and release them into the bloodstream for other cells in the body to use them as building blocks.
- Or it might store them inside the liver for a while.

Isn't that amazing, that some of the food you eat is used as building blocks to make bones, red blood cells, muscles, and so on? But no matter what you eat, all your food first goes to the liver. Only after processing does your broken-down food enter the general blood circulation.

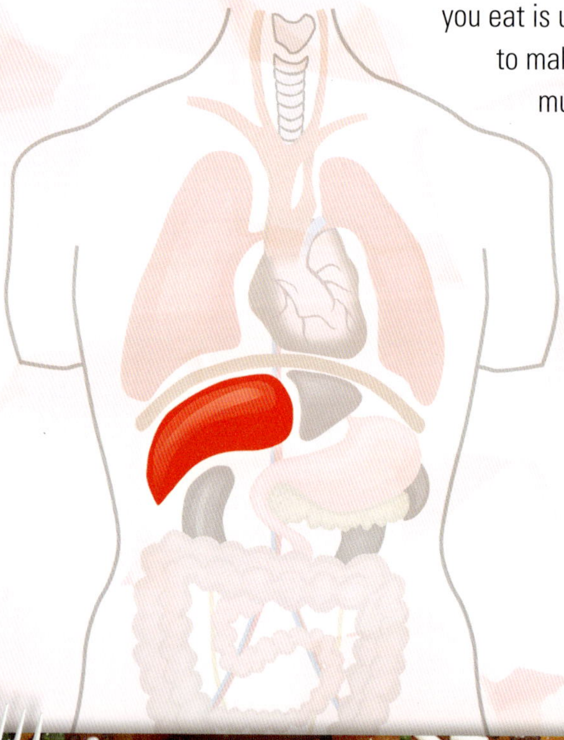

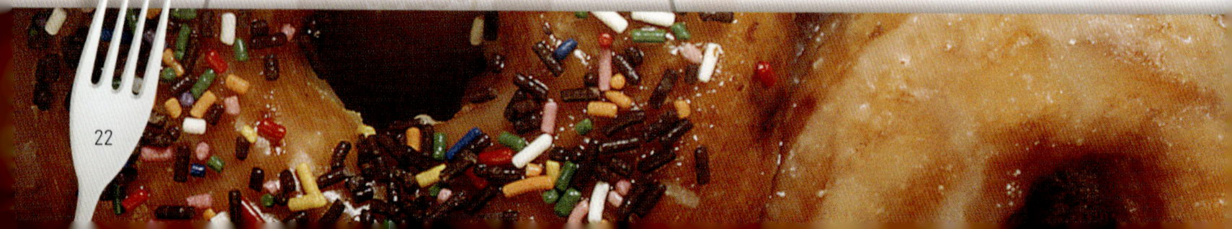

Food into Energy

In your food, every bond that joins one atom to another has energy in it. The job of your gut is to take this energy, and give it to your body.

In the case of a sugar, when the atoms are separated from each other, the energy is taken from the bonds that originally linked the atoms together. One reason for eating food is to get this energy. Another reason is to break large groups of atoms into building blocks that can be used to make other bigger chemicals.

To make energy, each carbon atom will be "burnt" with two oxygen atoms to make carbon dioxide, which is blown out of your mouth. Yes, your mouth is a major excretory organ.

But what happens to this "energy" that was taken out of the chemical bonds in the food of your meal? How is it used or stored?

In the short term, the energy is used to make one of the most important chemicals in the body: Adenosine Tri Phosphate, or ATP. This chemical is the energy currency of the body. To store energy, you make ATP. To use energy, you break down ATP.

In our society, whenever you want to do something, you have to spend money. In your body, whenever you want to do something, you have to spend ATP. If you shiver to keep warm, or move a muscle, or make an enzyme, or make your 1 litre of saliva every day, you spend ATP.

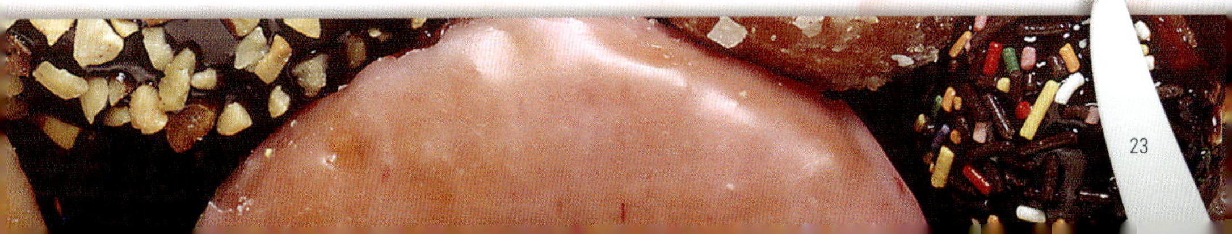

At any given instant, the total mass of ATP in your body is about 250 grams. But over the course of a 24-hour day, your body makes, and destroys, your own weight in ATP.

In ATP, there are three phosphates.

The third phosphate is stuck on very lightly. It needs only a tiny amount of energy to knock it off – but when it's knocked off, it releases a huge amount of energy. The Adenosine Tri Phosphate is converted down into Adenosine Di Phosphate. Then more energy, from your food, is pumped in to convert it back up to Adenosine Tri Phosphate. This happens to each molecule of ATP about 1000 to 1500 times each day.

So that's a short and really simplified version of what happens when we eat. Seems a lot of messing around, when from a mathematical point of view, all we have is the shape of a doughnut – a lot of stuff surrounding a hole in the middle.

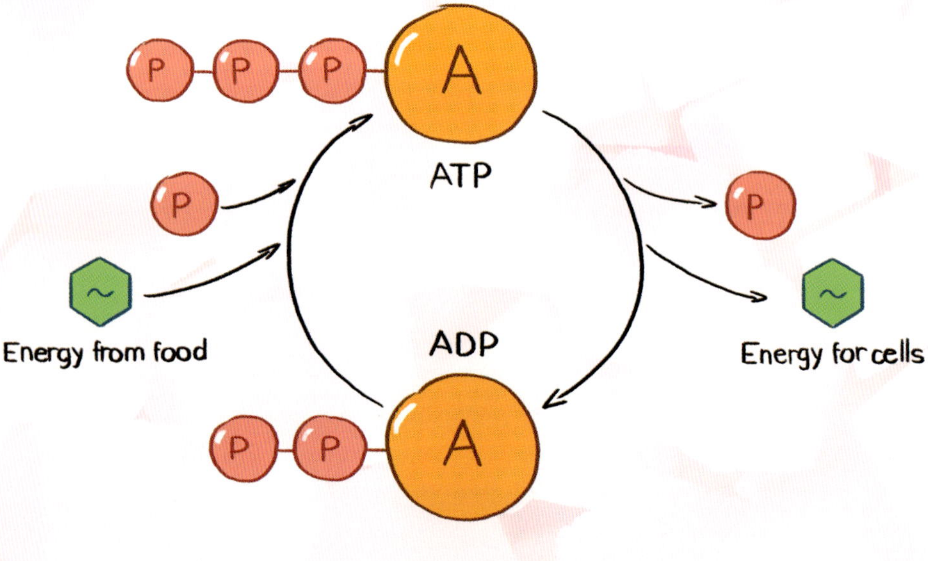

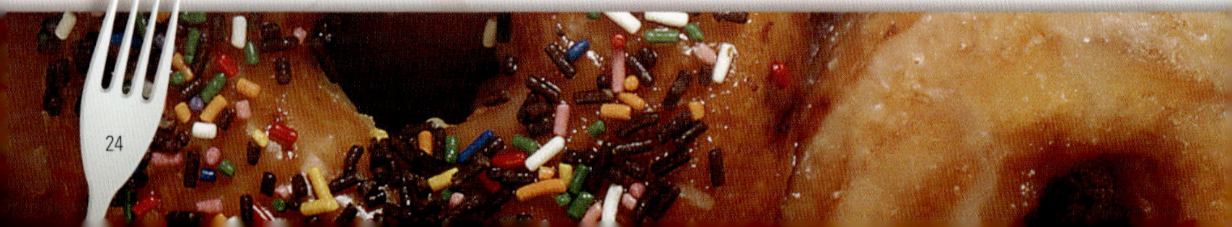

Up Close and Personal

find the answer on page 229

Beer - Balm for Burns?

Any burn that affects more than 20 per cent of a person's body surface area is classed as severe. Severe burns are medically really challenging to manage. Besides treating the actual burns and the potential infections, you also have to carefully judge how much extra fluid the patient will need to replace the fluid they've lost through their burnt skin. This is generally done via an intravenous drip.

But not always . . .

A 65-year-old Australian male who "fell into a garden fire", and burned 40 per cent of his body in the process, decided to wait till the next day before going to hospital. He didn't think the hospital would be open after hours. In the meantime, he treated himself by drinking six cans of San Miguel beer.

Incredibly, when he eventually did go to hospital, he was okay – despite receiving around 5 litres less fluid than he would have if he'd gone to hospital straightaway.

In terms of DIY, this man was very lucky indeed. Nobody seriously recommends resuscitation for burns with beer!

Perhaps his survival had something to do with San Miguel (St Michael) being the Patron Saint of Paramedics?

Up Close and Personal

find the answer on page 230

The Fat Virus

You've probably heard how powerful Social Networks are. They are so strong that you can even "catch" being overweight from your fat friends. You know the scene – you hang with them, you eat what they eat and then you get fat, too.

Well, it seems as though "catching" being fat might *literally* be true. There is a specific virus that can infect you and result in you gaining weight, because of the virus infection. But here's the kicker – while externally you look fat, internally your blood biochemistry says that you are thin!

Virus Makes Chickens Fat, then Dead

Let me emphasise that this research is in its Early Days. And catching this virus is a different kettle of fish from living your entire life with the "regular" microbes in your gut.

So while this theory of Weight Gain caused by a Virus is not mainline medicine, it's also not crazy. And it almost certainly does not account for most cases of obesity. It was put forward by Dr Nikhil Dhurandhar, Head of the Department of Viruses and Obesity at the Pennington Biomedical Research Center in Louisiana.

Dr Dhurandhar's interest began in 1988 when he was a junior doctor in Mumbai, India. A family friend, Dr SM Ajinkya, described to him how he had discovered an unusual virus known as SMAM-1. This virus had killed many thousands of chickens. As an Infectious Diseases Officer might expect, autopsies of the infected chickens showed an atrophied thymus, and pale and enlarged kidneys and liver. But what was unusual was that the dead chickens had excess fat in the abdomen. Surely, thought Dr Dhurandhar, if a chicken died of infection, having wasted away, it should be *less* fat, not *more*.

He began his research with chickens – 10 chickens infected with SMAM-1, and 10 not infected. He fed them equal amounts of food. Surprisingly, only the chickens infected with SMAM-1 got fat. Even more surprisingly, considering their excess fat, the levels of cholesterol and triglycerides in their blood plummeted. In you and me, if we gain weight our cholesterol and triglyceride levels usually rise.

Dr Dhurandhar then moved on to humans. He checked the blood from 52 of his overweight patients. About 20 per cent of his overweight patients had previously been infected with the SMAM-1 virus. These people weighed 15 kilograms *more* than the other overweight patients who had never been infected with the SMAM-1 virus. And, just like the chickens, the previously infected patients had lower triglyceride and cholesterol levels, even though they were more overweight!

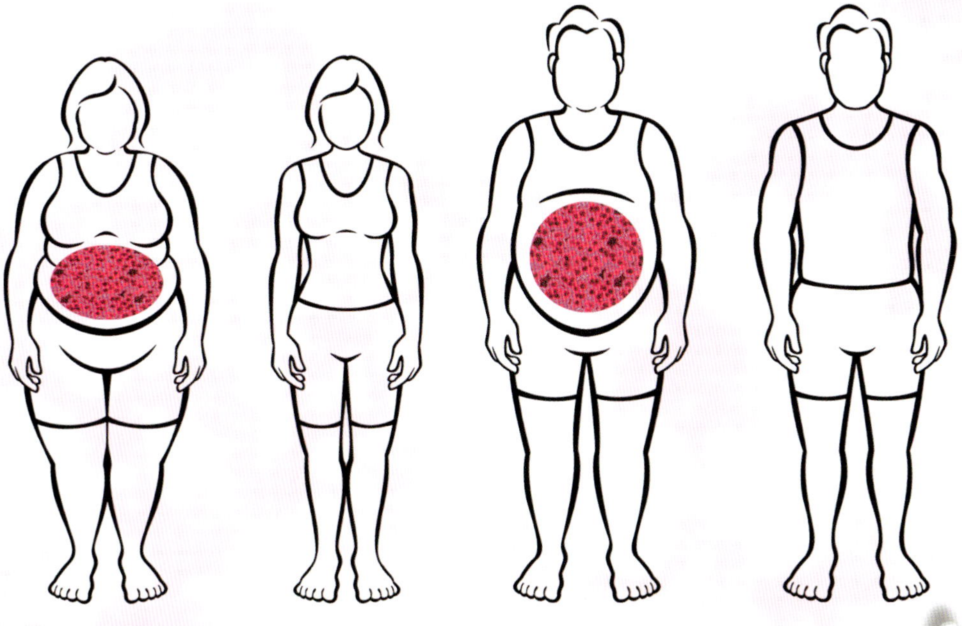

Human Virus Makes you Fat?

Dr Dhurandhar moved to the US to continue his research and in 1994 began working with Dr Richard Atkinson, then at the University of Wisconsin in Madison. Unfortunately for their research, the SMAM-1 Poultry Virus existed only in India, and US government regulations forbade its importation.

So they decided to work on a similar virus – a Human Adenovirus. In general, Human Adenoviruses cause symptoms that are so minor and so transient that most people don't even remember that they were ever infected. If they do remember, they usually think that they had a mild cold or an upset tummy. In the catalogue of the laboratory supplier there were 50 different Human Adenoviruses that the doctors could order. (Just like tools or undies, you can buy viruses out of a catalogue.) With amazing good luck, the first one they ordered – Adenovirus AD-36 – turned out to cause obesity in both animals and humans.

They started working with animals. They squirted Adenovirus AD-36 up the noses of chickens, rats and marmosets. Each of the species got fat and, with more time, obese.

They then moved on to humans. They could not deliberately infect volunteers with Adenovirus AD-36 to see if (as they predicted) the people got fat, while their cholesterol and triglyceride levels fell – that would have been unethical. So they invented a blood test that would identify people who had been previously infected with Adenovirus AD-36. They tested 502 volunteers – 360 obese and 142 not.

While a small percentage (11 per cent) of the non-obese people had been infected, a much larger percentage (30 per cent) of the obese people had been. And the previously infected people had the same paradoxical finding again – they simultaneously weighed more and had lower triglyceride and cholesterol blood levels than expected.

Careful Science

Dr Dhurandhar and Dr Atkinson did a little more checking.

Was it possible that obese people were more likely to get infected, for example, by Adenovirus? They checked if the obese and non-obese people had been previously infected by other strains of the Adenovirus. They had not. This suggested that they were not more prone to infection.

Was it possible that the combination of "obesity" plus "low cholesterol and triglyceride levels" was caused by natural variations in the human DNA? The doctors looked at 90 pairs of identical twins. In 20 of these pairs of twins, one twin had been exposed to Adenovirus AD-36 while the other twin had not. The previously infected twin carried an extra 2 per cent of body fat – even though they had identical DNA.

Further research has shown the Human Adenovirus AD-36 seems to both increase the number of fat cells in the body, as well as their size. Since then, another nine infective agents have been found that will make you fatter.

As I said at the beginning, it's still Early Days in this line of research, and the results are still tentative. But if this virus does cause obesity, you almost certainly won't catch it from an overweight person. You would most likely catch it from someone who has just been infected and is more contagious, and has not yet experienced the weight gain.

Today's medicine has only a small range of anti-viral drugs. So even if a virus infection can cause weight gain, you can probably manage the extra kilograms if you exercise and watch what you eat.

Which is lucky: reality TV shows might lose ratings as fast as their participants shed kilos if all they have to do is swallow a tablet, rather than generate sweat, blood and tears.

Shaken, not Stirred – Martinis are Forever

The James Bond books were written by British author Ian Fleming back in the 1950s and '60s. They were soon turned into the James Bond movies and have become an enduring institution. But the British Secret Intelligence Service agent (007) is not only indestructible, and not only does he know everything worth knowing at an international level . . . he also seems to have deep medical insights.

James Bond always orders his martinis "shaken, not stirred".

Why is this so?

Is Alcohol Good?

Professors Hirst and Trevithick from the Department of Biochemistry at the University of Western Ontario in Canada decided to find out.

It's safe to say that a little (let me emphasise "a little") alcohol is good for you, and might even reduce your risk of having cardiovascular disease, stroke and cataracts. Sure enough, James Bond, Secret Agent 007, has none of these medical conditions.

We don't know why a little alcohol is good for you, but it could be because of the antioxidant actions, or because of other chemicals present, such as flavonoids or polyphenols.

Professors Hirst and Trevithick compared the antioxidant activity of shaken and stirred martinis. They made mini-martinis by mixing 6 millilitres of gin with 3 millilitres of vermouth. These were either shaken (in a 100-millilitre medicine bottle for one minute) or stirred (in a 20-millilitre glass vial, using a vortex mixer).

Martini = Antioxidant?

First, it seems that "oxidants" in certain parts of your body are bad. Second, some antioxidants are good – for example, if they're in vegetables that you eat. But third, antioxidants that you buy in bottles probably do you no good, and might even be bad.

Anyhow, Professors Hirst and Trevithick measured how good a martini was at breaking down oxidants. (We are working on the assumption that oxidants are "bad", which is a little simplistic, but then so are the James Bond movies.) An easily accessible oxidant is hydrogen peroxide, H_2O_2. It's similar to Water – H_2O – but it has an extra oxygen atom. Thanks to this extra oxygen molecule, it's full of oxidant activity. So the scientists measured how well the different martinis deactivated hydrogen peroxide.

They discovered that gin by itself left a whopping 58.3 per cent of the hydrogen peroxide behind – which was pretty hopeless. Vermouth had a lot more antioxidant activity – it left behind 1.9 per cent of the hydrogen peroxide.

Stirred martinis were even better – removing all but 0.157 per cent of the hydrogen peroxide. But best of all were the shaken martinis, which left a microscopic 0.072 per cent of the peroxide behind.

The conclusion of the professors of biochemistry was obvious: "007's profound state of health may be due, at least in part, to compliant bartenders."

Talented Bartenders

Some talented bartenders study the specific gravity (or density) of various liquors. They then use this knowledge to gently pour different types of alcohol on top of one another so that the layers don't mix, making fancy drinks that seem to defy gravity.

The three-layered B-52 cocktail is usually served in a shot glass. The bartender first pours a coffee liqueur, such as Tia Maria or Kahlúa, then Bailey's Irish Cream slowly over the back of a spoon. Last, Grand Marnier is poured on top, again with a spoon.

And, with a flourish, you're served your pretty, layered drink.

Rage Against Research

Professors Hirst and Trevithick published their research on martinis and antioxidants in a letter to the Christmas 1999 issue of the *British Medical Journal*. This is traditionally the "fun" issue for the year, where all kinds of dodgy, tongue-in-cheek "research" gets published. Well, their innocent letter unleashed a flood of replies – many abusive!

Kathleen O'Malley pointed out that the real James Bond always drank a vodka martini, never a gin martini.

Dag Rekve from Norway, a Master of Business and Economics, also ticked off the biochemists for not knowing that Bond drank a vodka martini, and then added some rather stiff comments about the ugly side of corporate sponsorship: "Smirnoff heavily sponsors the Bond movies and the Canadian brewery Labatt funded the research and Corbott Distillers provided the liquor . . ." "Who," he went on to ask, were "the interests behind the advocacy of the possible health benefits of alcohol?"

Dr Shimon Barak criticised the biochemists for ignoring the fact that 007 would sometimes dip a smoked onion in his martini. Dr Barak pointed out that onions have an antioxidant oil (Allium cepa Linn) that "may help to explain the resistance of Mr Bond's organism to his addictions to cigarettes". Dr Barak also alerted us to the potential health benefits of the olives that Commander Bond sometimes consumed.

Wrong Bond?

While a shaken martini has more antioxidant activity than a stirred one, it is *not* your classic martini. Commander Bond was being his own man and was not following the proper recipe. It seems that while 007 is a darn fine secret agent, in the Land of Cocktails he is a dubious authority.

Andy del Rosal, a beverage chemist from the Bacardi–Martini Product Development Division (so he should know), pointed out that "Bond could be seen as somewhat of a heretic because the Classic Martini is stirred, not shaken, because shaking bruises the gin".

Indeed, Otto Andresen explained how this came to be when he noted that "the preference for shaken not stirred is a fault introduced by the manuscript writer for the first movie".

And finally, Professor Bell, the Head of Physiology from Trinity College in Dublin, also agreed that a martini should be stirred not shaken – when it is shaken, it becomes a drink called a "Bradford". Martini connoisseurs agree that shaking a martini makes it too "sharp".

Bruising?

What is this "bruising" that gin is prone to when it is shaken? I've never seen it come up in a dark colour. And I had no idea that gin had some kind of cardiovascular system.

Or has the gin been *emotionally* bruised?

It is claimed that when very cold vermouth (which is *not* gin) is shaken, some of the chemicals that were previously dissolved in solution take on a new form as droplets – giving the martini a hazy or cloudy appearance.

A different explanation is that shaking shatters the ice into tiny particles, and that these particles make the martini hazy or cloudy. A third and completely different explanation is that shaking the martini introduces a multitude of tiny air bubbles and, once again, a hazy or cloudy appearance.

However, bartenders tend to agree that shaking is essential for making cocktails that have ingredients that don't easily mix or dissolve in alcohol, such as eggs and dairy products.

But Professor Bell did congratulate the biochemists on discovering that "a mixture of gin and vermouth has far more antioxidant activity than either alone, and that this activity becomes mysteriously more powerful after shaking".

And he also gently pointed out that the biochemists were testing the wrong drink. In the early Bond classic, *Dr. No,* Ian Fleming specified Commander Bond's martini as six parts gin; two parts vodka; and one part vermouth. And the vermouth was a "kina lillet", which contains a lot of quinine, as well as various other herbs. In fact, Professor Bell says "the biochemists may have underestimated the medicinal power of a true Bond martini".

Up Close
and Personal
find the answer on
page 230

Cats Eschew Sweetness

Cats lap their milk and lick their fur – but you can forget about them accepting a sweetener to seal the deal.

It turns out that cats don't have a functioning taste receptor for sugar – even though the rest of a cat's taste buds are very similar to those of other mammals.

Back in the 1970s, scientists realised that cats were indifferent to sweet flavours. A bowl of water with added sugar was no more attractive to a cat than a bowl of unsweetened water.

But why? Mammals need two genes to make the taste receptor for sugar. Studies in various cats (tigers, cheetahs and domestic cats) showed that one of these genes has mutated and no longer works. As a result, cats can't make a functioning sweet receptor and they can't taste sweet things. (The sweet receptor is still present on the cat's tongue – but it is misshapen and doesn't work properly.)

Cats are most definitely meat eaters, and do just fine on a diet of protein and fat. But they don't eat carbohydrates. In evolutionary terms, there was no evolutionary pressure for cats to have a functioning sense of "sweet taste".

So while it might not be fair to call your cat a sourpuss, they definitely don't have a sweet tongue – or should that be tooth?

Olfactory Bulb
Nasal Cavity
Nostril
Hard Palate
Soft Palate

Taste that Smell

Supposedly, we humans have five senses – sight, hearing, taste, smell and touch. In fact, we have a whole bunch more – we can sense temperature, we have a sense of balance, we feel pain, we know where our arms and legs are and how fast they're moving, we understand time, and so on.

But let's ignore these other senses and stick to just two of the Traditional Five – "smell" and "taste".

There are a couple of common mythconceptions about these two senses. One has it that our sense of smell is so critically linked to our sense of taste that if you cannot smell what you are eating, you will have exactly the same taste experience every time – if the food items have a similar texture. For example, if you're blindfolded and asked to taste an apple, a raw potato, an onion and a carrot they'll all supposedly taste the same.

The other completely contrary mythconception would have us believe that smell and taste are totally independent of each other, and as such that smell has nothing to do with the enjoyment of a meal.

The truth is somewhere in between.

Just Test it!

Why don't people test things for themselves, instead of just repeating (and believing) outlandish claims?

For example, when the claim blossomed forth on the Interweb that mobile phone radiation was so dangerous and powerful that it could cook an egg in 10 minutes, my daughter Lola and I decided to test the claim ourselves. We Did The Experiment. We placed an egg between our two mobile phones and switched them on. After 60 minutes the temperature had not budged by one-tenth of a degree!

So back to our first mythconception. To do this test I pinched my nose (so there was no sense of smell) and closed my eyes (so I would not know what I was given). I could easily distinguish between an apple (sugary, flaky), a raw potato (bland, crunchy), a raw onion (pungent, obvious multiple-nested-layers texture) and a raw carrot (slightly sweet, lumpy texture).

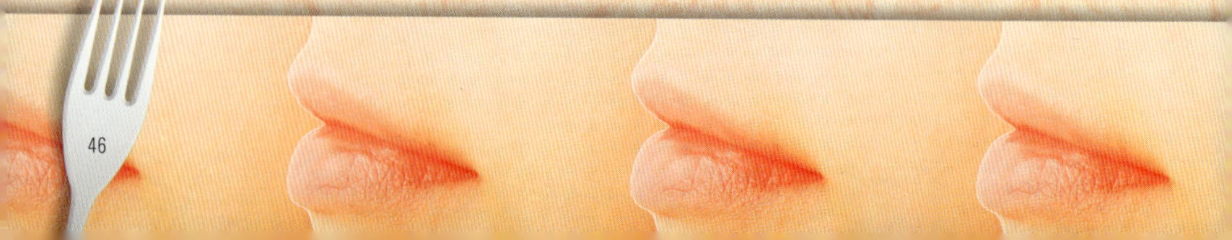

Taste

The sense of taste should be straightforward.

We put food into our mouth, mash it up with our teeth and swallow it. The liquid mush runs across our tongue and excites the taste receptors – and, a moment later, our brain registers the taste.

But in fact taste is much more complicated than that.

We actually begin to taste in the uterus at around the 14-week mark. Dr Beauchamp of the Monell Chemical Senses Center in Philadelphia says that a human foetus in its amniotic fluid (basically, its own urine) is continuously inhaling through its mouth. This amniotic fluid simultaneously tastes sweet (glucose), salty (sodium) and bitter (urine).

We have lots of taste buds on our tongue – but there are also some scattered around the inside of our mouth. Currently, we believe that the tastes we can sense are salt, sweet, sour and bitter. Recently, scientists in the field have agreed that there is another basic taste called "umami". Others claim that there is yet another one called "astringent".

Smell 101

The vastly under-recognised part of how we taste things is how we smell them. You've probably had the experience of not appreciating your food when your nose was blocked thanks to a cold. Dr Tracy Hollowood from the Sensory Science Centre at the University of Nottingham points out what we've all noticed at one time or another: that temporarily losing our sense of smell virtually destroys our enjoyment of food.

So why don't we normally associate smell with appreciating food?

Historical reasons, for one thing. Two millennia ago Aristotle relegated smell to the bottom rung of the ladder. He classed it as the most "undistinguished" of our senses. Then Immanuel Kant, the philosopher, wrote that our sense of smell was the "least rewarding and most easily dispensable" sense, and its only use was to warn us of toxic and poisonous foods. Even today, while countless poems and songs are written in praise of the mouth and eyes, the nose is mostly thought of as being held to the grindstone.

To smell "anything", molecules of that "anything" have to physically leave it and land on a postage stamp-sized patch of wet yellow tissue (olfactory epithelium) high in your nose. Here there are some five million neurons, each carrying only one type of receptor or sensor. Each receptor can react to several "smell" molecules, if they are closely related.

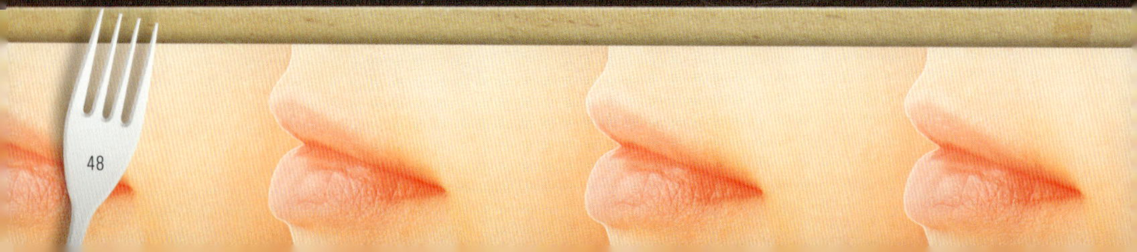

Train your Smell

We all know that you can train and build up your muscles. Surprisingly, you can also build up your sense of smell.

Dr Li and colleagues did an experiment. First, they tested how good volunteers were at differentiating odours. Second, the volunteers were exposed to just one of two odours for 3.5 minutes so they became very familiar with it. Finally, Dr Li and colleagues re-tested the volunteers with the first set of odours.

This is what they found: if the volunteer had been exposed to (for example) a floral odour for about 3.5 minutes, on re-testing they could select and differentiate the floral odour much better than when they were first tested.

So being exposed to one particular odour made them better at finding similar odours.

The Moral of this story is simple – use it or lose it!

When you sniff deeply through your nose, you take in more air – more molecules land on your postage stamp-sized olfactory epithelium and you can smell better for a few seconds. There are approximately 30 different kinds of "smell" molecules but we can potentially smell 10,000 different odours, depending on how many of each of these 30-odd smell molecules land on the olfactory epithelium. (Language works in a similar way, in that we can recognise thousands of different words that are made from just a hundred-or-so basic sounds.)

So when you enter a kitchen and "smell" a delicious meal, by this very act you have already taken some of the molecules from that meal into your body and started eating. In other words, smelling equals "micro eating".

Smell – Advanced

Even in the 21st century, our understanding of smell is still poor.

It was only as recently as 2004 that the Nobel Prize in Medicine or Physiology was awarded to Linda Buck and Richard Axel for their work in understanding this sense. Surprisingly, "smell" is the only sensory system to be completely mapped by scientists. Linda Buck and Richard Axel showed that about 1000 different genes in our DNA produce about 1000 different types of smell receptors or sensors.

While all 1000 smell sensors actively work in rats, only about 350 of them work in humans. Surprisingly, a lot of human DNA (a whole 3 per cent) is used to "encode" or "make" our smell receptors. Early in our evolution, our sense of smell must have been very important to our survival. Today, it's not quite as important. French perfume makers might disagree; and so might some French florists, too, who sell flowers arranged by fragrance, not colour.

Each smell sensor can respond to just a few different molecules. Most smells are made up of several different molecules. In any given smell, the various smell sensors send patterns of information to the brain and suddenly we recognise the smell of chalk dust, or a rose, or a curry.

So, when we think we are simply tasting, we are also, in fact, smelling . . .

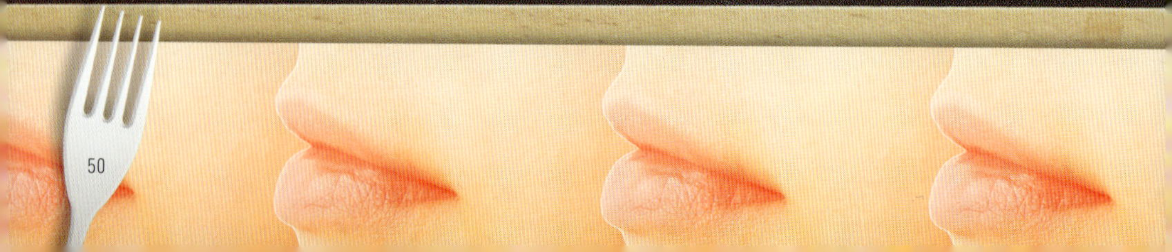

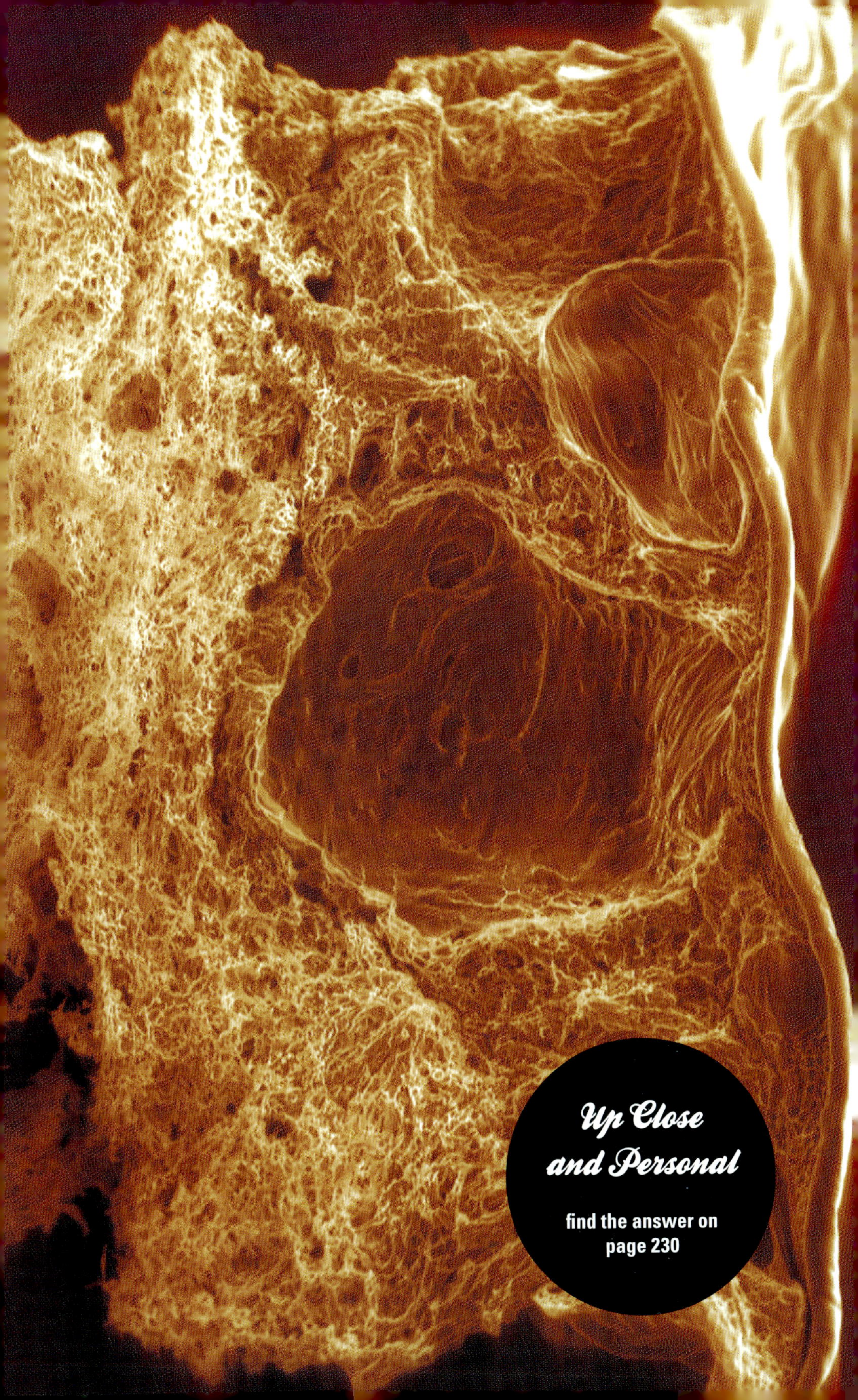

Up Close and Personal

find the answer on page 230

Antibiotics and the Gut

We know that the average human gut hosts about 1.5 kilograms of microbes – and that most of these microbes are bacteria. We also know that you carry in your bowels about 10 times more bacterial cells than the total number of your cells in your whole body.

But what happens to all these bacteria when you take antibiotics?

While we're on the subject of bacteria in the gut, here's something truly weird: human babies have a sterile, microbe-free gut – only after birth do they get colonised by bacteria and other microbes. But it gets even weirder: chickens do it the other way round – first they have bacteria growing in their gut, and then they hatch out of the egg.

Sterile Chicken Gut?

We know this weird fact about chickens because Dr Adriana Pedroso, from the University of Georgia, went to the trouble of incubating more than 300 eggs. She sterilised all the shells with a bleach solution and then carefully extracted the embryos using sterile techniques and sterile tools. Her DNA analysis of the intestines of the embryos found flourishing and diverse communities of microbes.

How did the microbes get inside the unhatched eggs? Well, the surface of the eggshell is porous. Dr Pedroso thinks that the microbes penetrated this porous surface before being swallowed by the developing embryo while it was still inside the eggshell.

The poultry industry has long fed chickens antibiotics to make them grow bigger and faster. Unfortunately, this seems to have caused an increase in nasty bacteria that are immune to these antibiotics. These bacteria have been constantly exposed to antibiotics over many generations, and so have evolved this resistance. And they've now got out into our community and are causing disease – and, in some terrible cases, killing people.

In the future, we may be able to establish a different community of healthy microbes in chickens before they're hatched. We might not need to feed them antibiotics any more.

Of course, we must acknowledge that part of the problem is Intensive Chicken Farming, where chickens are forced to live very close together in conditions they're just not used to.

Antibiotics and Gut Bacteria

Bacteria don't love each other just because they are bacteria. Sometimes they fight – usually with chemicals. Bacteria have already given us many useful antibiotics. Perhaps we can discover new antibiotics from the microbes in our gut?

As our knowledge increases and our techniques improve, we could also go further than simply killing bacteria with antibiotics. One day we might be able to develop medicines that mess with how bacteria "talk" to each other.

We know that bacteria live closely with other species of bacteria, and that they can "sense" the presence of these other species. We also know that they can release chemicals to inhibit or promote the growth of these other species. This is already the basis of another line of research for potential drug discoveries from the gut.

Antibiotics in Humans

Antibiotics have been around for nearly three-quarters of a century. When used to kill bad bacteria, they do a pretty good job. They produce remarkably few side effects, and these effects are vastly offset by the benefits. But bearing in mind that our gut is chock-a-block with bacteria, and that antibiotics are used to kill bad bacteria, we need to ask what antibiotics do to all the good bacteria.

First, most broad-spectrum antibiotics (which attack a wide range of bacteria) kill lots of your gut bacteria. These antibiotics can reduce the bacteria numbers (and mass) by a factor of 250–1000. So your normal 1.5 kilograms of gut bacteria could temporarily drop to just a few grams. But narrow-spectrum antibiotics (which attack only a small range of bacteria), such as vancomycin, hardly reduce the numbers at all. They do reduce the diversity of gut microbes, however. (Vancomycin is not a first-line drug, by the way, meaning it is usually administered only in hospitals and as a last resort).

Second, different antibiotics kill different microbes. In one study the broad-spectrum antibiotic ciprofloxacin changed the make-up of the various communities of bacteria in the gut. The effects were noticed within three to four days of taking the drug and were profound. However, within a week of finishing the one-week course the gut microbiome had begun to return to its original state. But the return journey does not always go all the way back to the original starting point. One patient permanently (or at least for the year-long duration of the study) lost, and did not regrow, one common genus of gut bacteria.

Third, some animals practise "corprophagy" – the eating of their own faeces. (For various reasons, this is definitely bad for humans and even a sign of madness, but it can be good for some animals.)

One study examined such animals that had been given antibiotics. As expected, the populations of their gut bacteria changed. But being able to eat the faeces of fellow animals that had not been given antibiotics helped their gut bacteria to recover back to normal.

In people whose gut microbiome has been permanently and badly changed by antibiotics, it is possible that a faeces transplant could help . . .

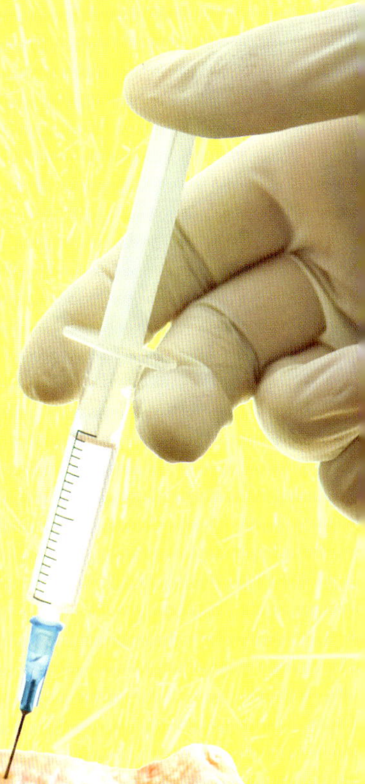

Could the Human Gut Make Antibiotics/IBD?

The cells that line our own gut are continually dying and being replaced – this happens to about 70 million of them every single day. Not only that, it also seems that our gut cells (as opposed to our gut microbes) make antibiotics. Furthermore, these antibiotics might be involved in that nasty illness Inflammatory Bowel Disease (IBD) – the common name for a group of diseases of the small intestine and colon. These include Crohn's Disease and Ulcerative Colitis. Symptoms range from vomiting and diarrhoea to abdominal pain and rectal bleeding.

Dr Hooper from the University of Texas Southwestern Medical Center did some experiments on germ-free mice with no bacteria in their gut. Dr Hooper introduced into the mice's guts different types of bacteria that naturally had sugars on the outside of their cell membranes.

The mice's gut lining recognised the bacteria as foreign and made a protein (a "lectin") that first joined onto the sugars on the bacteria's cell membranes, and then quickly destroyed the bacteria. Dr Hooper called them "killer proteins with a sweet tooth" – because they love sugars, and they love killing!

What does this have to do with IBD? Amazingly, similar lectins are found – in high levels – in patients with IBD.

Most of us get along just fine with the microbes in our gut. Perhaps, in people with IBD, their gut/immune system makes lectins that attack and kill some of the normal gut microbes. And perhaps this is what then leads to the painful symptoms of IBD. If so, we might be able to use this knowledge to "tweak" how, in some IBD sufferers, the gut and the immune system work together. The result could be to reduce the symptoms or – and here's hoping – cure the disease totally.

Big Finish

Your gut microbiome is one of the most dynamic ecosystems on the planet. It is also essential to our good health. Its composition changes day to day, but it does have a stable average state. Taking antibiotics can shift the average to another stable state. Will this be better or worse? We don't know – yet.

On the one hand, antibiotics do a lot of good. Since the first one, Prontosil, came onto the market in the early 1930s, we have found that the benefits vastly outweigh the side effects. I have had three cases of rapidly spreading cellulitis – a nasty infection of the superficial layers of the skin. I've had it on both hands, and on my right foot (from working on a car's brakes, another car's engine, and accidentally brushing my foot against a cloth-covered sofa and not even breaking the skin).

There are three basic treatments for rapidly spreading cellulitis: amputation, death or antibiotics. In my three encounters, the infection was spreading so quickly that recovery was not an option. Thanks to antibiotics, I'm still alive and I still have my four limbs. And, thanks to my lucky gut, I have never had any gut problems, or any other side effects, from taking antibiotics. But I am in a lucky minority: many people do suffer gut problems (tummy pain, diarrhoea and so on).

Each time you take a course of antibiotics, you are in a kind of lottery. The next roll of the dice could be the time that you lose an entire genus of bacteria in your gut and shift your microbiome to a new stable state – which could be worse.

The simple Take-home Message is: don't take antibiotics unless you really, really need them. And conversely, do take them if you do need them.

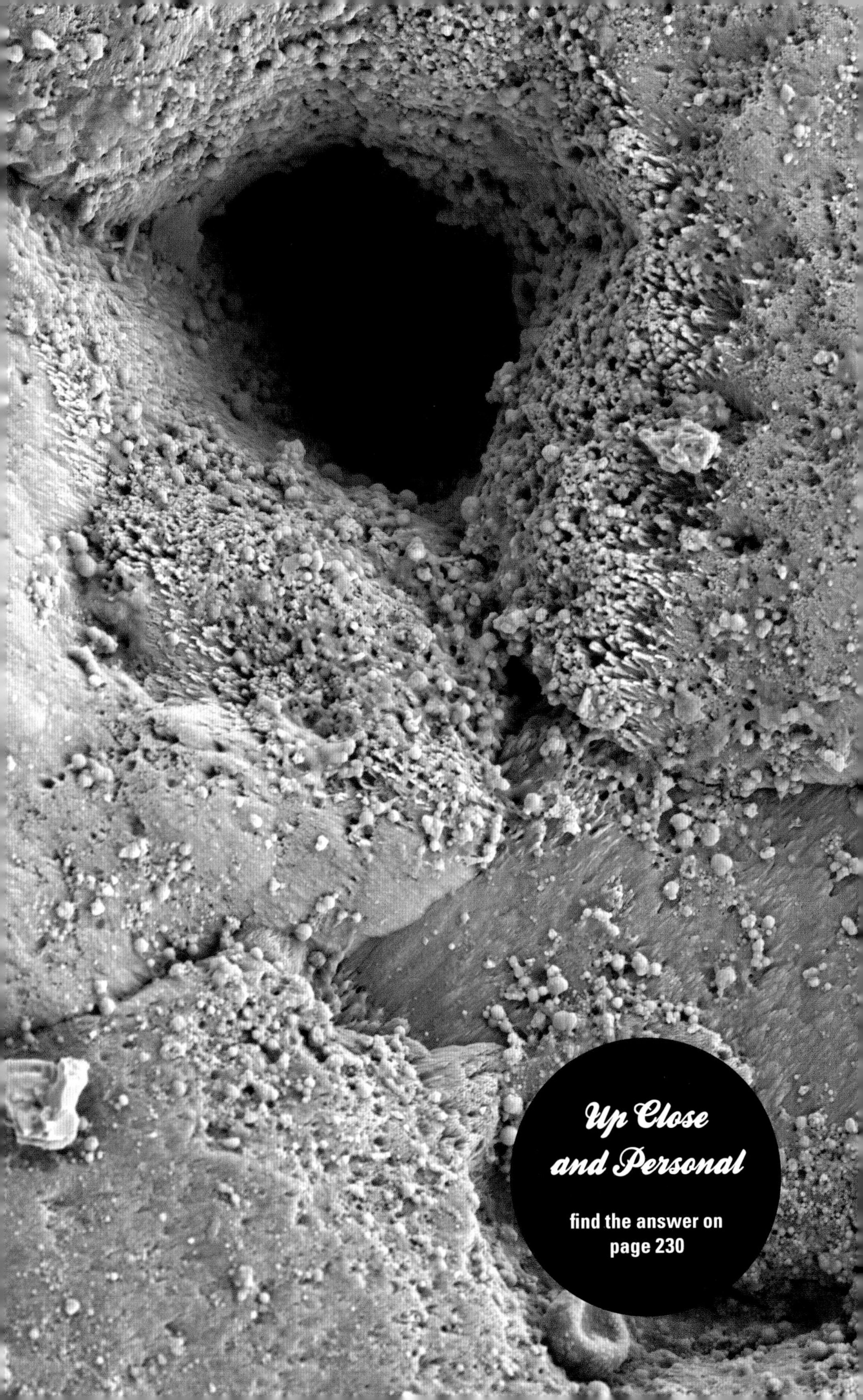
Up Close and Personal
find the answer on page 230

SHINKANSEN Series E3
SHINKANSEN Series E3

Popcorn and Pop Stars

Apparently, Madonna (the adored Pop Diva, not the Venerated Mother of God) claims that when she was a poor struggling artist in New York City (long before she became rich and famous) she lived entirely on popcorn. Is this nutritionally possible?

Popcorn 101

An unpopped corn kernel (or seed) is roughly the size of a pea. The vast majority (about 75 per cent) of the kernel consists of carbohydrate (mostly starch), with water making up another 15–20 per cent, and fat, proteins and minerals making up the rest.

When you heat the kernel, it "pops" and becomes an amazing 30–45 times *bigger*. Just 50 grams of corn can expand to about 2 litres of popcorn.

If you use air (not butter, oil or shortening) to heat and pop your 50 grams of corn, its energy content is about 800 kJ – roughly one-tenth of your daily energy needs.

But if you heat the corn with some kind of fat, it becomes one of the fattiest foods known to the human race. How come? Well, the popped corn has a huge surface area and it just loves to absorb stuff.

In one American study, researchers noted that a "large [serve of] popcorn had about 80 grams of fat, more than 50 of them saturated. That's almost three days' worth of saturated fat, or what you'd get from six Big Macs. And that's if you skip the [extra] 'butter' [that gets layered on top]".

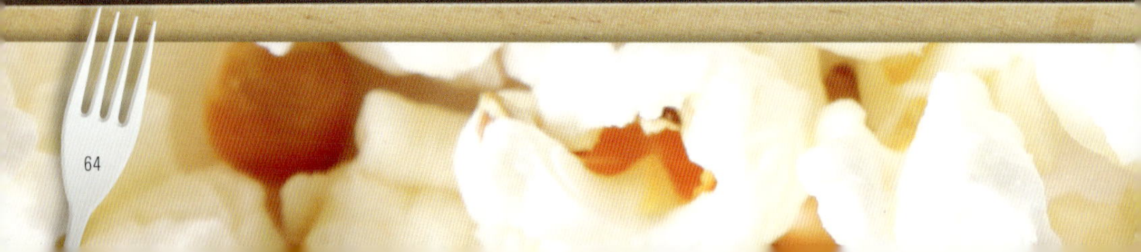

Bigger Corn?

Daniel Hong, a physicist at Lehigh University in Bethlehem, Pennsylvania, came across a problem plaguing a food company – how to make popcorn bigger.

Physicists say that physics can be applied to anything. Dr Hong knew that corn kernels burst only when the pressure inside the kernel is greater than the Burst Strength of the kernel's pericarp (or outer coating). He used concepts of Specific Heats, Adiabatic Expansion, Perturbations in Three Dimensions around a Sphere, and Time Dependent Equations of Motion to work out that to increase the volume of the popped corn by "X", he simply had to reduce the pressure inside the cooker (relative to the pressure inside the kernel) by "X times 1.3".
Easy!

How to Pop

Popcorn has been around for a very long time and has been found in thousand-year-old graves in places as far apart as Argentina and Mexico. Back then, across middle and upper South America, it seems the locals had a Popcorn Era.

Today, there are many steps on the Pathway to Perfect Popcorn.

First, you have to prepare the corn kernel.

Corn is dried on the cob out in the field until the water concentration inside the kernel drops to about 16 per cent. Then it's harvested and dried some more, until the water concentration drops to around 13–14 per cent. The higher-quality brands of popcorn are sold in airtight containers to keep their moisture levels around that wonderful 13–14 per cent mark.

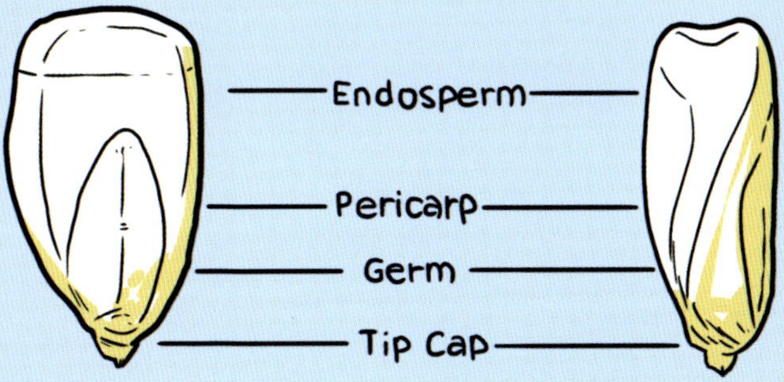

Once you get your slightly dried cob of corn, you can pull off one of the little kernels. The kernel has what's known as a "Tip Cap", which originally connected the kernel to the cob. It's shaped a bit like a squashed ball. The strong outer coating is called the "pericarp". It's actually *really* strong, and can resist a pressure of 9 atmospheres, or about 90 tonnes/square metre. Inside the pericarp is the "endosperm", which is mainly starch.

If you were to plant that kernel in the soil, the starch would be the early source of nutrition to power the growth of the "germ" (another part of the kernel) into a complete corn plant.

But instead of letting that starch become a food supply for the growing plant, let's make it "pop".

Explosions

If you piled up gunpowder and lit it, all you'd get is lots of smoke and heat.

But if you make a firecracker by wrapping the gunpowder in lots of layers of strong paper and then light it, you'll get an explosion.

As the gunpowder burns, the volume of gas inside the firecracker increases very quickly – and so does the pressure. But the gas is trapped. When the pressure gets great enough to overcome the strength of the paper wrapping, it bursts through suddenly – and you have an explosion.

A similar process happens when you pop a corn kernel.

First, you need to apply heat to the kernel. Most of the moisture is in the starch. At 100°C the water turns into steam and expands. Once the pressure reaches around 90 tonnes/square metre, it ruptures the skin (or pericarp) of the corn kernel with a sudden pop. The combination of the heat and the sudden expansion turns the previously compact starch into soft, fluffy popcorn.

Madonna Lives

And what about Madonna? According to Professor Jennie Brand-Miller from the University of Sydney, the "pop" princess could not have survived for very long living on popcorn alone. To get the nutrients essential for life, she must have been eating a little meat (or a non-animal source of well-balanced protein) and milk.

The Insurmountable Problem is that the protein in corn contains very low amounts of the amino acid tryptophan. Tryptophan is essential for making niacin, Vitamin B3. Corn does contain a tiny bit of niacin, but it is bound up to other chemicals and so is biologically unavailable to the human body.

In order to survive on popcorn alone, Madonna must have had an extra source of niacin. Niacin is found in red meat, fish, poultry and green leafy vegetables. Or maybe she did what the people of South America (where corn originated) did: they treated the corn with lime before turning it into a tortilla.

The alkalinity of the lime released the bound niacin, preventing Popcorn Pellagra (see overleaf), which otherwise would have appeared within a few weeks.

So Madonna either cracked the Big Time in under two weeks, or she was popping vitamins along with her popcorn . . .

Fortnight of Corn = Gone

The terrible disease "Pellagra" was first described in Spain in 1735. It was endemic around the Mediterranean (in Italy, Spain and elsewhere), places where corn and maize had been introduced and become the dominant food crop. At the time it seemed reasonable to believe that there was some kind of toxin in corn that was causing the disease.

Between 1906 and 1940 pellagra spread, attacking some three million Americans (mostly in the South), around 100,000 of whom died. Many of these people ate only the "Three Ms" – Meat (actually pork back fat), Molasses (mostly sugar) and Meal (cornmeal) – and it was this limited diet that gave them pellagra. In turn, the pellagra caused the "Four Ds" – Dermatitis, Diarrhoea, Dementia and Death. Indeed, the name "pellagra" comes from the Italian words "*pelle*" meaning "skin", and "*agra*" meaning "sour".

In 1914 the American Congress set up the Spartanburg Pellagra Hospital in South Carolina to study the disease. By 1915 Dr Joseph Goldberger had proved that pellagra was not caused by a toxin – with an elegant, if "messy", two-part study.

In Part 1 of his study, Dr Goldberger and his wife and assistants carried out what they charmingly called "Filth Parties". They collected the blood and other bodily fluids, scabs and even faeces from patients with pellagra, and ate or injected themselves with this "filth".

Thankfully, nobody got the disease. No disease meant no toxin.

In Part 2 of his study, Dr Goldberger worked with 11 "volunteers" from a Mississippi prison, all of whom had been promised a full pardon in exchange for their cooperation. First, he made sure they were healthy by giving them clean clothes, accommodating them in houses that were cleaned every day, and feeding them a balanced diet.

He then changed their diet, feeding them only corn. Within a fortnight the prisoners began to complain of mental confusion, loss of appetite and headaches. By the third week, 7 of the 11 prisoners began to suffer full clinical pellagra.

When Dr Goldberger put them back on a balanced diet, the pellagra vanished.

Dr Goldberger proved that corn did not contain a toxin but, rather, that it was missing something important. He later showed that brewer's yeast (which we now know is rich in B vitamins) prevented pellagra in a corn-only diet. But he had not identified corn's missing element.

Happily, in 1937 Tom Spies, Marion Blankenhorn and Clark Cooper discovered that niacin was the missing vitamin, and that replacing it could cure pellagra in humans.

Popcorn Worker's Lung

Many workplaces have their own special diseases. Generally speaking, the more specialised the work, the more specialised the disease.

Back in 2001, "Popcorn Worker's Lung" was noted at a factory that made Microwave Popcorn. The official name of the disease is "Bronchiolitis Obliterans Syndrome". (By the way, try to avoid getting any disease that has the word "Obliterans" in its official name. You really don't want any of your body parts "obliterated", especially if they are the smaller airways known as "bronchioles".)

It took until 2007 to discover that the culprit causing the Bronchiolitis Obliterans Syndrome was a chemical called "diacetyl". Diacetyl is found naturally in milk products such as milk, butter and cheese. It's also used as a flavouring ingredient in the synthetic "butter" in Microwave Popcorn. When heated, diacetyl forms a vapour. If you happen to inhale this vapour, it scars and damages the small airways in your lungs. Sufferers can breathe in deeply, but have trouble breathing out.

It turned out the vapour was present in high levels inside the factory.

Surprisingly, a consumer who bought Microwave Popcorn from a shop was also diagnosed with Bronchiolitis Obliterans Syndrome. How could he have gotten enough exposure just by making popcorn at home?

Well, not only had this man eaten Microwave Popcorn at least twice a day for 10 years, every time he made it he would deeply inhale the vapour from the bag once the hot steam had evaporated. Every day he was breathing in heated diacetyl vapour, which is where the highest risk lies.

So he quit. And six months after giving up Microwave Popcorn, he'd lost over 20 kilograms in weight and his lung function had begun to improve.

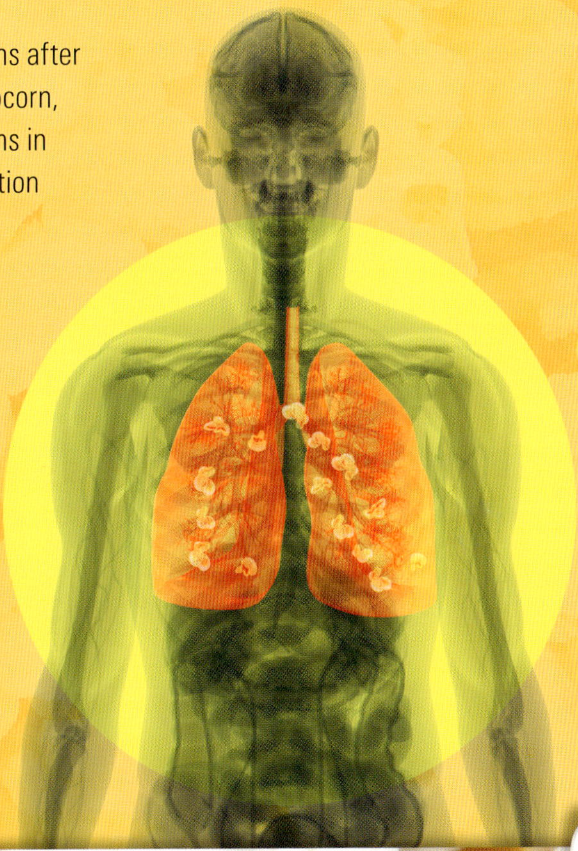

The Strangers Within

When you look at yourself in the mirror, it seems reasonable to assume that what you see is mostly "you". Sure, there might be some bacteria living on your skin, maybe a temporary flea that jumped off a passing dog, perhaps even some lice from the regular and very annoying primary-school infestations . . . But, by and large, "you" should be mostly "you".

But it ain't necessarily so, bro. We are less human than we think – about 90 per cent less. Yep, thanks to Evolution, 10 per cent of our cells are human, and 90 per cent are microbes. (I am talking about the number of cells, not their weight.) In fact, if an Alien looked at you, they would see "you" as just a carrier for all the other life forms.

And these life forms mostly live in your gut (or bowel).

What is a "Gene"?

The word "gene" comes from the Greek words "*genetive*" and "*genesis*", meaning "origin", or how things began.

In the First Meaning of Gene (from way back in Ye Olde Days), people spoke of the gene for blossom colour, or for height, and so on. That was because of Gregor Mendel, an Austrian Augustinian monk and scientist. In the mid-1800s he bred and tested some 29,000 pea plants. These pea plants had blossoms that were either purple or white. He spent over a decade on this single experiment, breeding for different coloured blossoms.

The Second Meaning of Gene (that is, the more recent meaning) is a section of the DNA that makes a single protein. According to this meaning, genes are responsible for biological traits such as eye colour or blood type.

But, of course, it's more complicated than that.

The Current Meaning of Gene is that it's a section of the DNA which is related to something that can be inherited. So a gene might be responsible (by itself) for regulating or controlling a biological trait; or it might share this duty with other sections/genes in the DNA. So several genes might be responsible for eye colour, and have different specific jobs.

I wonder what next year's Meaning will be?

You = Ecosystem

You are never alone.

Your microbes are with you from the moment you are born to after you die, when they recycle you.

Yep, we humans are mostly made from other life forms. We are a super-organism – a blending of us and them.

Remember that only about 10 per cent of the cells in your body came from that single fertilised egg that was made when your mother and father loved each other very much in a very special way . . . There are about 10 trillion of "your" cells. These carry the DNA beloved by the Forensic Crime TV shows – the DNA that will identify "you" as uniquely "you".

Of the other 90 per cent of the cells in your body (about 100 trillion) that belong to other living creatures, the vast majority are microbes living in your gut. A very small percentage live in your stomach, a lot more in your small intestine, but most live in your large intestine. They make up about half the weight of the contents of your bowels – or about 1.5 kilograms.

These microbial communities are called the human "microbiome" (from "*bios*", the Greek word for "life"). They are also called your "microflora" and your "normal flora".

The result is that each of us is a strange microbe–human hybrid. Yes, we are more Microbe than Man.

Classifying Life

There are a few different ways to classify "Life", but a recent method splits it into three separate "Domains" – Eukarya, Bacteria and Archaea. (Just to make things easy, let's leave "viruses" out of this discussion.)

First, a little background knowledge. Living creatures are made from cells. Sometimes, the entire living creature is just one single cell, for example, a bacterium. But sometimes the living creature is made up of tens, or thousands, or millions, or billions, or trillions, or quadrillions of cells, such as a whale.

Second, more background knowledge. Cells use DNA as their "blueprint". Living creatures store their DNA inside their cells. When cells split into two identical copies, a "biological machine" inside the cell "reads" this "blueprint".

One Domain of life is "Eukarya". The word comes from the Greek words "*eu*" meaning "good" and "*karyon*" meaning "nut", "core" or "kernel". These creatures have their DNA wrapped up inside a membrane. Everything inside the membrane is called the "nucleus". Eukaryotes developed around 1.6–2.1 billion years ago.

We humans are eukaryotes, as are all the other animals, plants and fungi. We think that we're the dominant life form on the planet, and are also pretty wonderful because we invented poetry, weapons of mass destruction and income tax.

But eukaryotes make up only a tiny fraction of all living creatures. In fact, you could easily fit all the humans on the planet into a cube, one kilometre on each side.

Another Domain is "Bacteria". The word comes from the Greek "*bakterion*", meaning "staff, rod or cane". This is because under a microscope the first bacteria discovered looked a bit like rods. In a bacterial cell the DNA is not wrapped up inside a membrane – instead, it floats freely inside the cell. Bacteria can exist as a separate individual single cell, or in clusters or groups. These groups can sometimes make decisions about their next actions, such as whether to migrate or to reproduce. Some bacteria can double their numbers every 10 minutes. Bacteria probably developed around four billion years ago.

Typically, there are over 100 million bacteria in each glass of clean, fresh drinking water, and 40 million in each gram of dirt. Only a microscopically tiny fraction of bacteria do harm to us humans. We "control" the "bad" bacteria with our immune system and occasional help from antibiotics.

Phylogenetic Tree of Life

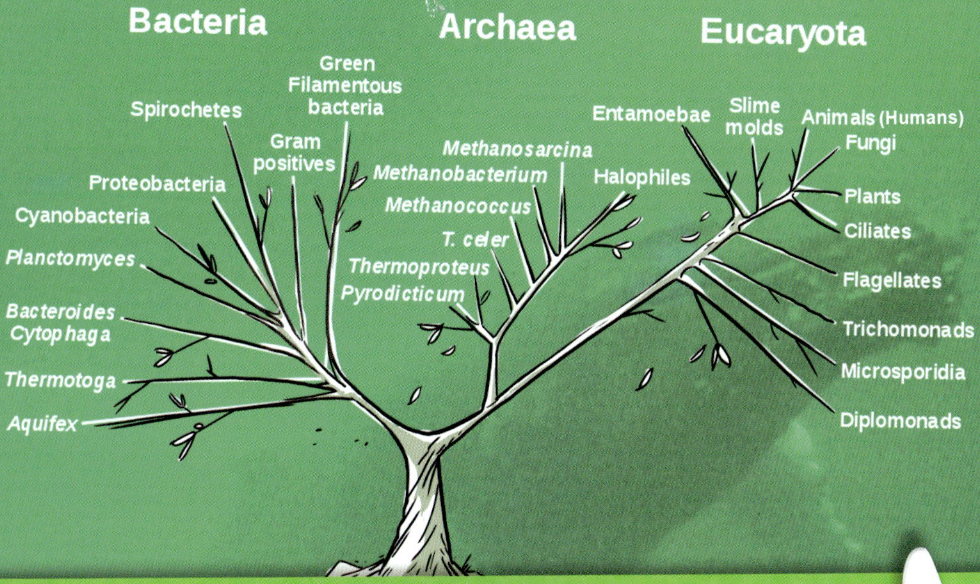

The Last Domain of life

The last Domain of life is *"Archaea"*. Their name comes from the Greek word "archaea", meaning "ancient things". They probably developed around the same time as bacteria. Like bacteria, archaea can live as individual cells or can group themselves into colonies, which seem to act and behave like a single living creature.

Archaea seem similar to bacteria because of their size and shape, and because they don't have a separate membrane inside the cell to wrap up the DNA. But archaea have a very different evolutionary history from bacteria, and often use quite different biochemistry and different metabolic pathways to make things happen. For example, the archaea cell membrane (which is like our human skin in that it's a barrier that keeps the outside outand the inside in) is very different from the bacterial cell membrane.

The first-discovered archaea were so-called "extremophiles" because they live in extreme environments. They were found in temperatures above 100°C in oil wells, geysers, and in the "black smokers" on the ocean floor that squirt out mineral-laden water at temperatures up to 464°C. They can also survive environments that are very cold, or very salty, or very acidic, or very alkaline. They even have group names such as "thermophiles", "halophiles", "acidophiles" and "alkaliphiles" to represent their extreme native environment. Since then, they have been found in all environments.

Microbiome and the Borg Collective

The microbial communities in our bodies (our gut microbiome) are a bit like the Borg Collective from Star Trek.

The Borg is an assembly of billions of electronically linked semi-synthetic creatures. Once, they were human or alien. But now they are all part of the Borg, which works as a single intelligence directed towards a single goal.

Similarly, our gut microbiome works towards survival of both itself – and us.

TRUTH
from the Mouths of Babes . . .

Surprisingly, a lot of the science in this story comes from nursery rhymes.

You remember how little girls are made of "sugar and spice and all things nice". This is correct, because they are indeed made from sugars, fats and proteins.

But, would you believe it, another line from the same rhyme is also kind of correct when it says that little boys are made of "slugs and snails and puppy dogs' tails". Yep, there are all kinds of critters in the DNA we carry.

Of course, in these Politically Correct Times, the Nursery Rhyme Gender Specificity is a Historic (and Inaccurate) Reference.

Popular Place?

It's surprising that our gut is such a popular place. We know that it's popular because it has a population of 100 trillion.

So why live there? On the downside, the microbes can be attacked by both the human immune system and antibiotics; they get pushed around by food and faeces; they can be eaten by other microbes, and they can get flushed into our toilet bowl.

But, on the upside, food and water is always flowing past them, the temperature is constant and they have lots of friends.

We ♥ Gut Microbes

We have a fair and reasonable relationship with our gut microbiome.

We give them both room and board in the shape of a lovely home in the comfort and safety of our human gut. Every time that we eat, they eat. They grab, store and redistribute energy from the food we put in front of them. They use this energy to maintain and repair themselves. They reproduce, and they communicate with each other.

In return, they do stuff for us. They process much of the food we eat to make it digestible. Energy salvage is one of their major jobs; in fact, they generate as much as 30 per cent of your daily energy. They also make vitamins for us and they protect us against nasty microbes – either by attacking them or by not allowing them to find a home in our gut. They are essential for the development of the immune system associated with the lining of our gut.

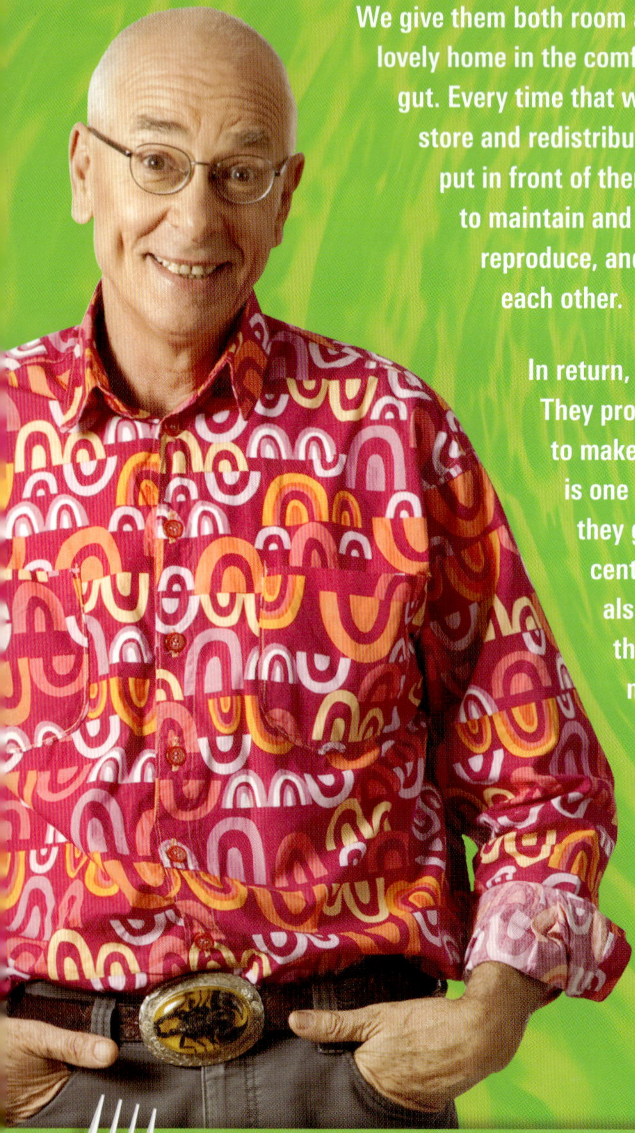

Our gut microbes help the cells that line our gut grow and proliferate and turn into different types of cells. They also help make the tiny capillaries that line the outside of the gut. These blood vessels pick up the proteins and carbohydrates from what we eat. Without them, our gut would be a much smaller shrivelled version ofits healthier and robust microbe-laden self.

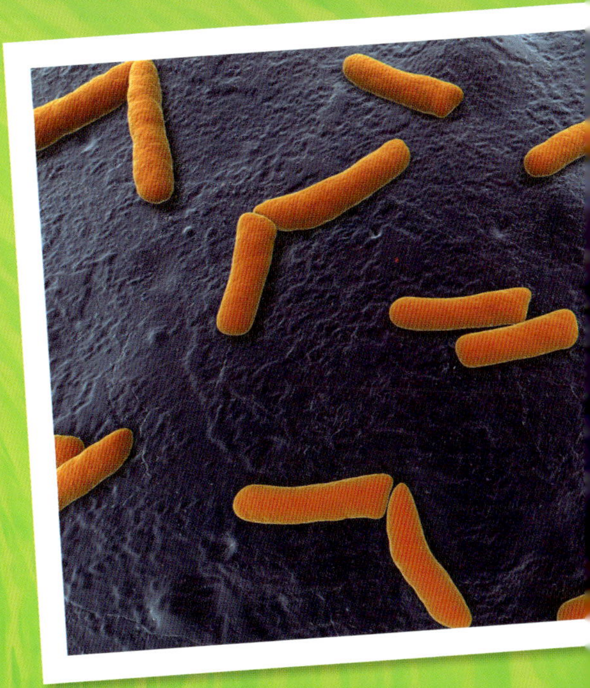

Microbes also break down carbohydrates that we cannot digest or extract energy from. For example, we humans have about 98 enzymes that break down carbohydrates. But one single bacterium (*Bacteroides thetaiotaomicron*) living in our gut has, by itself, over 460 enzymes (and other chemicals) to turn carbohydrates into energy. These extra enzymes can break down sugars (pectin, sorbitol and so forth) that our human enzymes cannot. Some of the energy goes to microbes, but the rest goes to us. In fact, if it weren't for the single-celled creatures in our gut, we'd all be a lot thinner. Microbes are best known for causing diseases, but this is a false impression which has arisen out of bad publicity. Most microbes are neither good nor bad to us, and only a tiny minority go out of their way to do bad things to us. Certainly, our gut microbes are very good to us.

As our understanding of the gut microbiome grows, we will find "biomarkers" to monitor our short- and long-term health, especially with regard to the food we eat. In time, we will better be able to "adjust" our individual, idiosyncratic ways of dealing with infections and orally delivered drugs and antibiotics.

Who Lives in your Gut

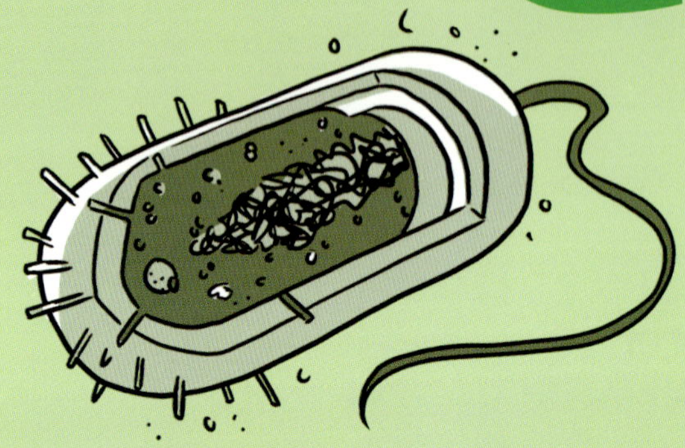

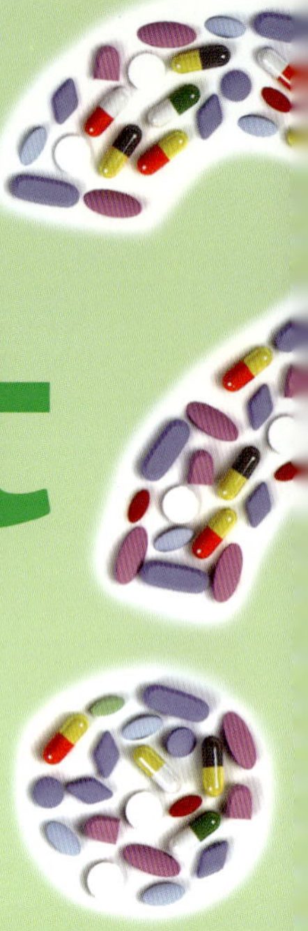

There are at least 1000 different species of single-celled microbes colonising your gut. Recent figures suggest there could be many thousands. They are overwhelmingly Bacteria. Bacteria break down sugars into short-chain fatty acids, carbon dioxide and hydrogen gas.

Only 8 of the 55 known "Divisions" of Bacteria have been found in the gut, but only 3 of them are common. In terms of numbers of cells, 2 bacteria (*Bacteroidetes* and *Firmicutes*) each make up about one-third of the microbes in your gut. The third one, Proteobacteria, trails behind.

The rest of the microbes are called Eukarya (fungi, protozoans and the like) and Archaea (mostly found in the large intestine, or distal gut). One group of the Archaea, the "methanogens" that live in the end gut of humans and herbivores, speed digestion of your food. They do this by taking in the hydrogen made by other microbes and turning it into the gas methane. (So, yes, it is possible to ignite your "fart gas", but don't do this experiment at home, folks.)

Beside the "regular" microbes (Bacteria, Eukarya and Archaea) that live in the human gut, about 1200 different viruses have also been found. Most of these are called "bacteriophages" because they "eat" or attack bacteria. About 80 per cent of these are unknown to science. There are probably 10 times as many Viruses in our microbiome than Bacteria. Viruses are the "dark matter" of the biosphere.

In terms of genes, Bacteria make up about 93 per cent of the genes found in the gut, Viruses about 6 per cent, Eukarya about 0.5 per cent, and Archaea about 0.8 per cent.

The combination of our gut and the microbes that inhabit it form what's known as an "ecosystem". In nature, we have different ecosystems such as tropical jungles, deserts, swamps, tundra and savannahs – and different creatures live in different ecosystems. Once things are stable, every single ecosystem is inhabited – but by different critters. In the same way, each person has different ecosystems in different parts of their gut. And the ecosystem at the same site in the gut (stomach, ascending colon and so on) is different from person to person.

We can classify humans into three main Classes, depending on what ratio of populations of gut bacteria they carry. (It's a bit like classifying people by Blood Type.) It's Early Days in this research (because we haven't yet checked out Chinese people and Africans, and so on) but initial research shows that one Class makes more of vitamins B7, B2, B5 and C, and gets its energy by fermenting carbohydrates and proteins. The second Class makes more of vitamins B1 and folate, and gets its energy from glycoproteins in the layer of mucus on the wall of the gut. The third Class is very good at breaking down mucins.

Drugs and food have to get past these different microbes before they can enter the bloodstream. Perhaps these different classes of microbes can help explain why different people absorb drugs and nutrients differently?

Marine Stowaway

Some of the genes in your gut arrived there by the most circuitous pathways – at least, if you're Japanese.

In the ocean, red algae contain specialised sugars that are not found in your average food plant on land. So several marine bacteria, including Zobellia galactanivorans, evolved enzymes to break down these sugars for energy and raw materials. (It's a bit like needing the right tool for the job, such as a tin opener opener to deal with canned food.) At this stage, biochemist Dr Jan-Hendrik Hehemann was not particularly surprised. After all, it would be silly to let a potential source of food be wasted.

So Dr Hehemann – just to be thorough – went looking for any genes that might make these very specific enzymes. Amazingly, he found them. But only in the guts of people living in Japan.

How did they get there?

One theory is that these genes hitched a lift on the *Zobellia galactanivorans* bacteria that travelled into the gut on uncooked seaweed.

The Japanese wrap sushi in uncooked seaweed called "nori". On average each Japanese person eats 14.2 grams of seaweed each day. In fact, Japanese records from the eighth century show taxes being paid in seaweed – so, way back then, seaweed was important to the Japanese culture and cuisine. The *Zobellia galactanivorans* bacteria survived the trip from ocean to human gut – two very different ecosystems – but once in the human gut, they died out. Before dying, they passed the genes to make this enzyme to another bacterium, *Bacteroides plebius*.

Zobellia galactanivorans no longer live in the Japanese gut, but their genes remain in the Japanese gut microbiome, breaking down uncommon sulphates for food and energy.

Strangely, in Dr Hehemann's study, one infant fed entirely on breast milk had the enzyme in her gut microbiome. Maybe it passed to her from her mother?

Another theory is that the genes were carried from an oceanic bacterium into the Japanese DNA by bacteriophages. Bacteriophages are a large group of viruses that infect bacteria. They are probably the most common biological creatures on our planet. And they are very good at transferring genes from one creature to another.

Today, our supermarkets offer us energy-rich processed foods that can be totally sterile without any microbes at all. You certainly don't want food poisoning, but maybe it's a good idea to occasionally go to foreign places and eat unfamiliar foods.

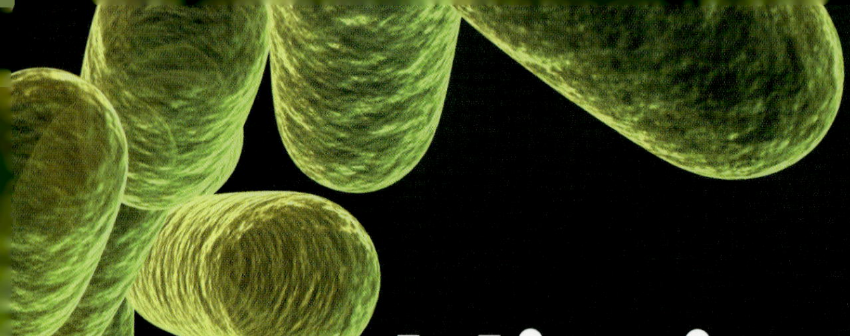

Missing Microbes

We have "known" for about a century that there is some kind of amazingly diverse and dense community of microbes in our gut. So how come it took us until the 21st century to begin to get some real understanding of what actually lives there? The answer is easy – we couldn't grow them.

Microbiology is the study of micro-organisms. In Ye Olde Days, the microbiologist would place a thin layer of sterile food on a sterile plate and smear some microbes onto it. The microbes would grow, and the microbiologist would examine them with a microscope and do a few other tests.

The first problem with getting the microbes in our gut to grow on a culture plate is that they exist in an oxygen-free environment, so oxygen will kill them. The second problem is that we don't know what they like to eat. The third problem is that some of these microbes are Bacteria, but some are not – they belong to different Domains such as Archaea and Eukarya. All these factors combined means that we can't grow over 99 per cent of the microbes that live in our gut.

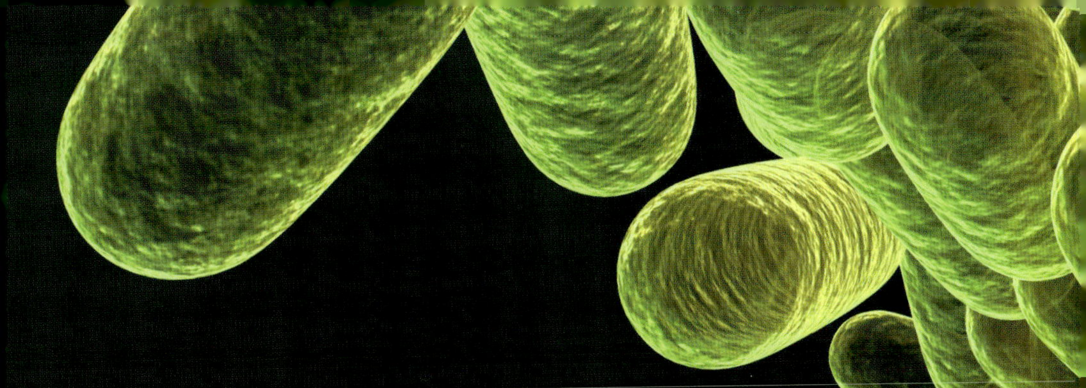

Luckily, Nobel Prize–winning research has given us a new technology to find microbes – Polymerase Chain-Reaction Amplification (PCR). PCR does not try to grow the microbes on a glass plate – it looks directly at the DNA. Before PCR, we humans had discovered some 5000 species of bacteria. Then PCR was used to look at what lived in a single teaspoon of dirt. Suddenly, another 5000 species of microbes were discovered!

Thanks to the new tool of PCR, it was now possible to start looking for microbes – both very easily and quickly, and on a large scale.

In 2006 the European Commission (with various research institutes) committed $US31 million to a project examining the gut microbiome – called MetaHIT (Metagonomics of Human Intestinal Tract). In 2007 the US National Institutes of Health approved a five-year, $US115 million plan to study microbes that live anywhere on our body – the Human Microbiome Project.

The Take-home Message

We humans are really just a long tube surrounded by "other stuff".

This Tube of Life – your gut – is surprisingly big. If you rolled it out it would be as long as a bus – about 10 metres long. If you flattened it longways and sideways, it's as big as a tennis court – about 200 square metres.

There are all kinds of critters living in your gut. Most of them live in the distal gut – the far end, or the large intestine. And the overwhelming majority seem to be bacteria. There are more of these microbes in your gut than there are stars in our Milky Way Galaxy. In numbers, these microbes make up over 90 per cent of the cells in your body. But because these microbial cells are so small, they weigh only 1.5 kilograms.

There are more than half a million microbial genes in an individual person's gut. But it takes only 23,000 genes to make a human. In term of genes, these microbes carry over 95 per cent of the genes in your body – so we are, in fact, less than 5 per cent human! (A study that checked the gut microbes in 124 people found a total of about 3.3 million different microbial genes.)

Without these critters you would be a sickly, skinny person with shrivelled and abnormally small internal organs, continuously eating in a futile attempt to put some flesh on your bones.

Up Close and Personal

find the answer on page 230

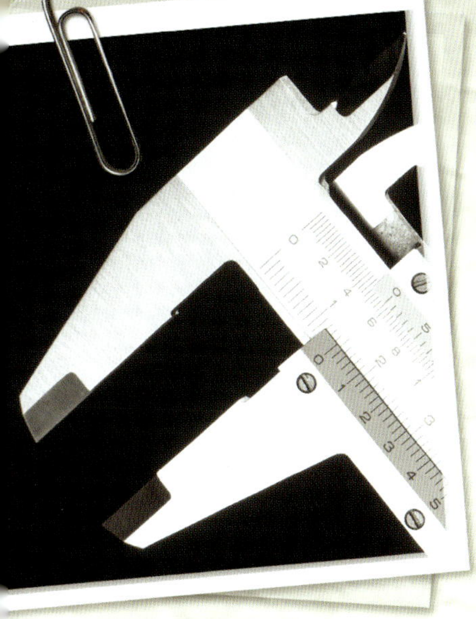

The Perfect Cheese Sandwich

By avoiding pre-sliced cheese, we claw back ownership of the thickness of cheese we cut for our sandwiches. But then we face the big question – how thick should the slice be?

The British Cheese Board commissioned the very same man who won an Ig Nobel prize for research into biscuit dunking, Len Fisher of Bristol University, to solve this vexing problem.

The goal was to measure the levels of different "cheese aromas" released by biting into slices of different thickness. Where in the body do you make this measurement? Obviously, at the back of the nose!

Dr Fisher used the MS-Nose, invented by Professor Andy Taylor from Nottingham University – which involved slipping a plastic sampling tube up one nostril until it reached the back of the nose.

The first finding was that as you increased the thickness of the cheese slice, the levels of "cheese aromas" also increased – but only up to a certain limit.

Second, this limit depended on the type of cheese – 2.5 millimetres for Double Gloucester/Red Leicester, 2.8 millimetres for Cheddar, 3.0 millimetres for Stilton, and 7.0 millimetres for Wensleydale. This thickness should be the same across the entire slice of the cheese.

Third, and rather surprisingly, you should spread butter on the bread. When the butter reaches the tongue, it tends to "hold" the "cheese aromas" for longer than if it were not present.

Finally, adding tomatoes improves the cheesiness, while pickles do not.

But I cannot see the link between this research and the Old Spanish Saying, "A cheese without rind is like a maiden without shame".

CHIPS

Fat Germs

When it comes to controlling our weight, the scientific breakthroughs made during the last five years are truly astonishing. Even so – and here's the catch – there are no Quick or Simple Fixes.

We now know that the type of microbes you have in your gut can determine whether you're fat or lean. Not only that, according to experiments done on mice this fatness/leanness is actually "transmissible" – that means it can be transmitted from a lean mouse to a fat mouse (and vice versa) just by swapping gut microbes.

And, even more strangely for women, we also now know that being overweight might be related to one single tiny microbe – that lives in your mouth.

Read on . . .

Energy In vs Energy Out vs Hormones

Here are the facts.

First, if your energy intake matches exactly the amount of energy you burn, your weight will remain stable.

Second, if you increase your energy intake over a long period of time – even if only by a tiny amount – and you don't compensate for this by burning more energy, your weight will increase.

Between the ages of 25 and 55, the average American person puts on around 0.45 kilograms of fat each year. This corresponds to an energy surplus of a tiny 50 kilojoules per day. Surprisingly, the excess energy is less than 1 per cent of the total diet, even for obese people. Even so, this tiny difference, accumulated over years, can lead to obesity.

The human body tries to buffer this change in weight in various ways. One way involves hormones. For example, if your fat levels are low, your levels of a hormone called "leptin" will also be low. In humans, weight loss leads to falling leptin levels, which in turn leads to more eating and less exercise, which in turn leads to getting fat again.

Of course, the interlocking and complex physiological loops that control our energy balance aren't run by one single hormone. Lots of hormones and nerves talk to, and listen to, our brain. There are hormonal systems that sense our energy stores and demands; systems that regulate how much we eat; systems that control how much of the food we eat is broken down and absorbed in our gut, and other systems that control how much energy we spend, and so on.

Obesity 101

In the US, about 30 per cent of the population is overweight. They have what is called a Body Mass Index (or BMI) of 25.0–29.9. Another 30 per cent of people are classified as obese (they have a BMI of more than 30).

According to the US Department of Health and Human Services, obesity kills around 300,000 Americans each year. It's the second biggest preventable killer after cigarettes. Around 15 million Americans are fat enough to qualify for weight-loss surgery, based on current guidelines.

Obesity rates in Western society have soared since World War II. The World Health Organization estimates that some 300 million people worldwide are obese. Obesity is a Big Global Problem because it can increase your risk of stroke, Type 2 diabetes, high blood pressure, sleep apnoea, osteoarthritis and many other nasty conditions.

Two causes of weight gain can be easily explained: the ready availability of cheap calories, and the reduction in physical exercise as part of our daily lives. You can control these to some degree. But there's other stuff that's not so easily controlled – including genetic, environmental and cultural factors.

“Natural” Human Obesity Genes

The first human gene that promotes weight gain was discovered in 1994. Since then, another 50 have been discovered. They interact with our energy intake and energy burning in many different ways.

Some genes help dedicated fidgeters use up as much energy as if they had run a marathon. Other genes “tell” you how much to eat, while others “tell” you when you’re full. Some genes regulate how much energy you turn into fat, some regulate where that fat will go (for example, to the hips, thighs, between shoulder blades or the bottom), and some regulate how long all this will take.

There is another important factor in the Obesity Epidemic over which you have little control: the types of microbes in your gut.

Back in Ye Olde Days, when food was hard to come by, having microbes in your gut was a Big Advantage. The little critters helped you extract extra calories from your food – and both you and the microbes benefitted. However, these days, at least in the West, food is easy to come by.

But what about those people who swear blind that they eat hardly anything and yet still put on weight?

Eat Less, Gain More

Janet S is a good example. She was a patient of Dr Richard Atkinson of Virginia Commonwealth University. Dr Atkinson has studied obesity for more than thirty years. He met Janet S in 1975 when he was at the University of California in Los Angeles. She was 25 years old and weighed 158 kilograms.

Dr Atkinson was studying 30 people who weighed more than 136 kilograms. Like most obese people, they all consumed more energy than they could burn. Each day their meals were giving them an average of 28,000 kJ – four times more than is needed by an average lean person!

The deal was that if the volunteers stayed in hospital for three months so the researchers could monitor them, they would each receive free surgery at the end – an intestinal bypass.

They were put on different diets and the changes in their weights were closely monitored.

But one volunteer – Janet S – was different.

The dietician calculated how much she had to eat to maintain her weight of 158 kilograms. She ate exactly what she was given – no more, no less – but even so, she gained about half a kilogram *each day*.

What was going on?

Microbes Extract Extra Energy from Food

Another study compared "microbe-free" mice with mice that had carried gut microbes since birth.

But hang on – what are "microbe-free mice" and how do you get them?

First, you very carefully deliver the baby mice by sterile Caesarean Section. Second, you raise these mice inside sterile plastic bubbles, with airlocks. And third, you feed them food that has no germs on it.

If you have been careful enough with your sterile techniques, you will now have germ-free mice. They'll have no germs on the surface of their skin on the outside, or on the surface of their gut on the inside. Unfortunately, they won't be very healthy and will have defective immune systems. And they are very different from "regular" mice that have lived with microbes all their lives.

Anyway, in the experiment, "regular" microbe-carrying mice ate 29 per cent *less* rodent chow. Amazingly, their bodies contained 42 per cent *more* fat.

Did the microbe-free mice stay thin because they had a higher metabolic rate (that is, did they burn up more energy at rest)? No, the exact opposite. They actually had a metabolic rate that was 27 per cent *lower*. The lack of gut microbes meant that they could not extract all the potential energy from their food.

In us humans, the microbes in our gut increase the amount of energy available to us by about 10–30 per cent. *Polysaccharides* are long chains of repeating identical sugars. They are the most common biological polymers on Earth – but without gut microbes, we cannot extract the energy they contain.

Fat Changes Microbe Levels

The microbes in our gut somehow "know" how much energy we eat – and adjust their population numbers to suit.

A neat little experiment back in 2006 proved this. The scientists (Drs Ley and Gordon and colleagues) examined 12 obese volunteers and 5 lean volunteers. The obese people had 20 per cent more *Firmicutes* and 90 per cent fewer *Bacteriodetes* than the lean people. (The way I remember this is "F" for "fat", "F" for "Firmicutes".)

The scientists put the obese people on a low-energy diet (low fat or low carbohydrate) and after one year these people lost as much as one-quarter of their body weight. At the same time, the ratio of *Firmicutes* to *Bacteroidetes* in their guts changed – they had fewer *Firmicutes* and more *Bacteroidetes*.

This showed that if you're lean, your microbe populations are different to when you were fat.

Costly Gut

Maintaining our gut "microbiome" is not free. It takes energy, and this energy comes from our food.

Each day, the average British person excretes about 15 to 20 grams (dry weight) of gut microbes. The energy to replace them comes from the digestion and processing of 50–65 grams of sugars.

But if your gut microbiome needs more energy than this, you will tend to be more lean.

Microbes Change Fat Levels

Dr Turnbaugh and his colleagues showed that changing microbe populations can make mice fat.

First, they transplanted microbes from a human gut into very lean microbe-free mice. These microbes grew and flourished in these so-called "humanised" mice – and the mice got fatter.

Second, they fed half of these mice a low-fat diet, and fed the other half a high-fat diet. As you would expect, the mice fed a high-fat diet became much fatter than the mice fed on a low-fat diet – and they were all fatter than the microbe-free mice.

No surprises there.

And just like the Drs Ley and Gordon study, the fatter mice had higher proportions of *Firmicutes* bacteria in their gut microbiome. Amazingly, the proportions changed in the astonishingly short time of just one day.

Third, they kept the diets the same but transplanted the microbes. Lean mice, now carrying the gut microbiome from the high-fat-fed "humanised" mice, quickly became obese. In other words, the "fat-adapted" gut microbes can – if transplanted – make you fat.

Perhaps this is how Janet S, Dr Atkinson's patient, was able to gain weight when she ate a supposedly "neutral-weight" diet. Perhaps her gut bacteria could extract lots of energy from what she ate.

Crazy Evolution?

But super-efficient microbes in obese people don't quite make sense – at least, from an evolutionary point of view.

When a person's energy intake matches exactly the amount of energy they burn, they can be lean. Their gut microbes are moderately efficient at extracting extra energy from what they find in the gut, and passing some of it on to their human host.

But when a person eats *more* than they burn up, they become fatter. Their gut microbes (and this is surprising) get very efficient at extracting even more energy from the food. So once you get fat, it's easy to put on more fat.

Why? They don't need to do this. Why save the maximum efficiency for when you don't need it? Surely it makes more sense to be super-efficient when there is hardly any food. Then, as more food becomes available, and there is no longer the urgency to avoid starvation, the microbes can become less efficient.

Instead, they do the exact opposite. As people lose weight, these microbes become less efficient. This means that it gets easier to lose weight. Maybe this is where "crash diets" could have a limited usefulness.

We still have no idea how the microbes "know" if you are skinny or lean.

"Misleading" Packaging Labels?

Packaged food generally comes with a label giving you an energy and nutrient breakdown. But what it can't tell you is how that food will be processed in your body. Even though 250 grams of milk might "contain" 680 kJ of energy, not everybody can extract all that energy.

First, highly efficient gut microbes, which are very good at extracting energy, could lead to obesity. And, of course, variations in the human DNA lead to variations in the gut microbiome.

Second, you might naturally have a slightly higher or lower rate of burning up energy (the BMR or Basal Metabolic Rate).

Third, it might take a really long time for food to pass through your gut. The longer the transit time, the more energy the microbiome can potentially extract.

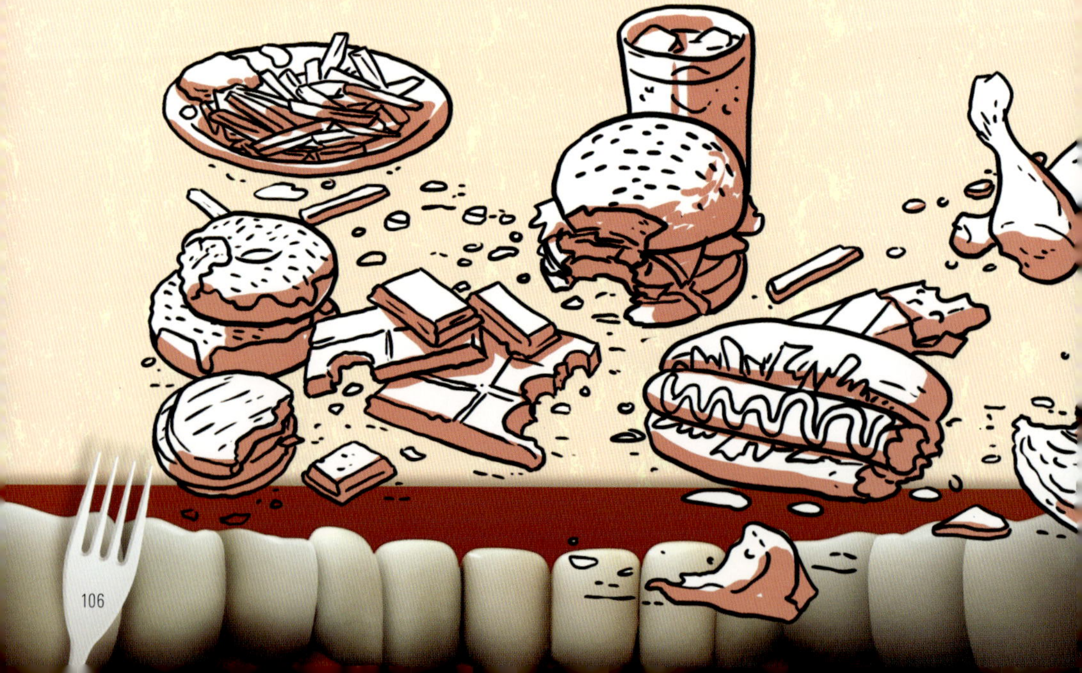

Mouth Bacteria and Obesity

A study by Dr Goodson compared the mouth microbes of 313 overweight women (all with a BMI of 27–32) with the mouth microbes of 232 average-weight women.

Over 98 per cent of the overweight women had a single specific species of bacteria: *Selenomonas noxia*. This bacterium accounted for more than 1 per cent of all the different bacteria in the women's saliva.

If you swallow a litre of saliva per day, you'll be swallowing one gram of bacteria in total – and about a billion cells of *Selenomonas noxia*!

The big unanswered question is this – did the bacteria cause the women to become overweight or vice versa? Or are the bacteria nothing to do with being overweight, and just an accidental association? We'll have to wait and see.

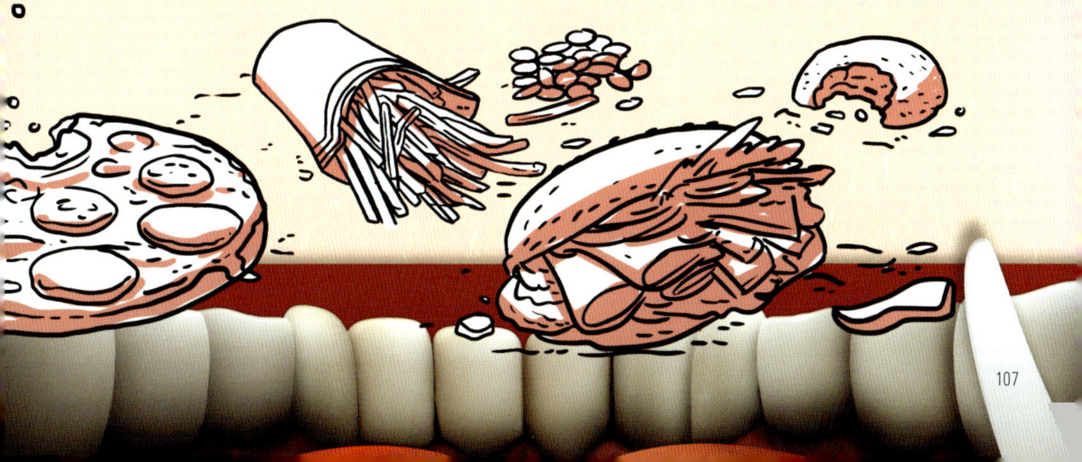

Marshmallow Moments

You usually think of food as having only a few purposes – easing hunger, building social bonds, and so on. But according to Walter Mischel, you can predict a person's future from how they deal with marshmallows when they are very young.

Back in the 1970s, Walter Mischel left a group of four-year-old kids alone in a room with a single marshmallow and a bell. The rule was that if they waited for him to come back they would get two marshmallows – but if they rang the bell to call him back they would get only the single marshmallow.

The kids lasted differing times – from less than a minute to more than fifteen minutes.

The kids who waited longer went on to do better at school (based on exam results). But the kids who waited the least time were "doomed" to be bullies, disappointments to their teachers and parents, and more likely to have drug problems.

It seems that being able to delay gratification is a useful skill for success in adult life.

Can self-control be taught? The answer is yes. It is possible to detect differences in ability to pay attention from as early as nine months of age.

The trick to resisting the temptation of the marshmallow lay in being able to distract yourself and think about other things. This knowledge should be a fundamental part of early education.

Up Close
and Personal
find the answer on
page 230

Panda's Puzzling Palate

In the overwhelming majority of cases, the 1.5 kilograms of microbes in your gut are your friends. When you were very little, they helped both your gut and your immune system develop fully. In fact, if you didn't have these microbes in your gut, you would have to eat a lot more, and yet you would be a lot more skinny. Mostly, you and your gut are perfectly adapted to each other.

But that does not seem to be the case with the black-and-white Giant Panda. The Giant Panda is a bear and so belongs to the Carnivore family. It has the gut of a meat eater, and it has the gut microbes of a meat eater. And yet its diet is 99 per cent vegetarian.

This simply doesn't make sense.

Carnivores vs Herbivores

Different diets lead to different types of guts and, of course, different populations of microbes that live in those guts.

Carnivores (meat eaters) tend to have a short gut. They're in the minority – only about 20 per cent of today's animals are carnivores. They have a narrower diversity of microbes in their gut, as compared to herbivores.

Herbivores (plant eaters) can survive in a much wider range of environments – if there's vegetation to eat, they can usually survive. They have a longer gut than carnivores. This is because the complex carbohydrates in plants are fairly resistant to being broken down. A longer gut means a longer transit time, so the microbes can better do their job of fermenting and breaking down carbohydrates.

Herbivore Group 1 do their fermenting in the foregut. These animals include kangaroos, sheep, cattle and giraffes. This means that as the partly digested food moves along the gut, the microbes travel with it. So the microbes get eaten by other microbes in the hindgut – it's a tough life.

Herbivore Group 2 expand their hindgut for fermenting. Horses, elephants and gorillas are part of this group. This means that as the digested food gets excreted into the outside world, so too do the microbes. So the microbes get "excreted" – and again, it's a tough life.

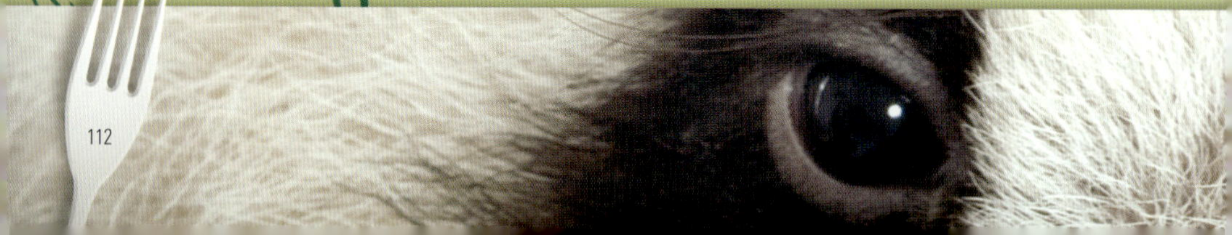

Black and White, yet Mysterious

The Giant Panda is very distinctive with its large round body, and the large black patches around its eyes and ears, and on its body. They are big animals – the males weigh up to 150 kilograms, the females up to 125 kilograms. There are fewer than 3000 Giant Pandas on the whole planet, so they are classified as an Endangered Species. This sad statistic, combined with the Giant Panda's very cute face, is probably why the WWF (World Wide Fund for Nature) chose it for their logo.

The Giant Panda is a fussy and finicky feeder. As the seasons roll by it chooses different species of bamboo, and even different parts of the bamboo, to eat. For example, in late spring and early summer it enjoys the shoots that are rich in sugars and starches. But for the rest of the year it eats the hard, indigestible stalk.

And now we run into the crazy mystery.

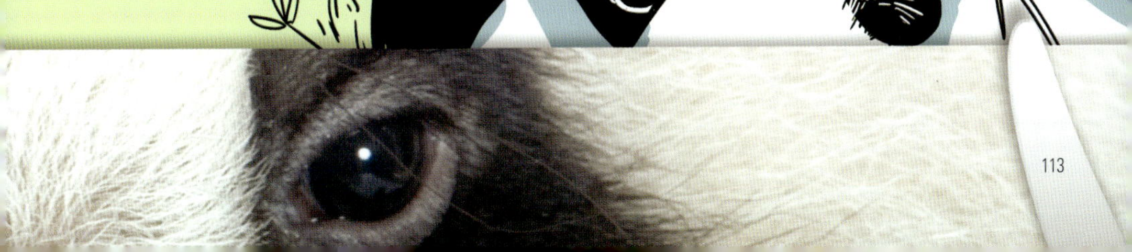

Paradoxical Panda

The vegetarian Giant Panda has the short gut of a meat eater – just a simple single stomach and straight colon. Unlike most of the other plant eaters, it doesn't have the multiple stomachs that act as digestion chambers – or fermentation vats – to break down the otherwise indigestible cellulose in its diet.

Furthermore, the DNA of the Giant Panda does not have the genes to make enzymes that can break down the cellulose that it eats. That wouldn't matter if it had the gut microbes to digest the cellulose – but it doesn't have them, either. Instead, it has microbes such as *E. coli* and *Streptococcus*.

As a result, the Giant Panda's entire life is based on continual eating to gather enough energy just to survive. It spends over 12 hours each and every day chewing up to 25 kilograms of bamboo. But most of the 25 kilograms of bamboo pass through the Giant Panda's gut undigested. And comes out in the Giant Panda's very frequent 40-or-so daily defecations.

In its relentless pursuit of food the Giant Panda hardly travels, and it definitely avoids other Pandas. In fact, possibly to save on energy, the female Giant Panda has a very short pregnancy and gives birth to a tiny baby weighing less than a cup of tea – about 140 grams. That really is tiny, considering that the mother weighs 125 kilograms.

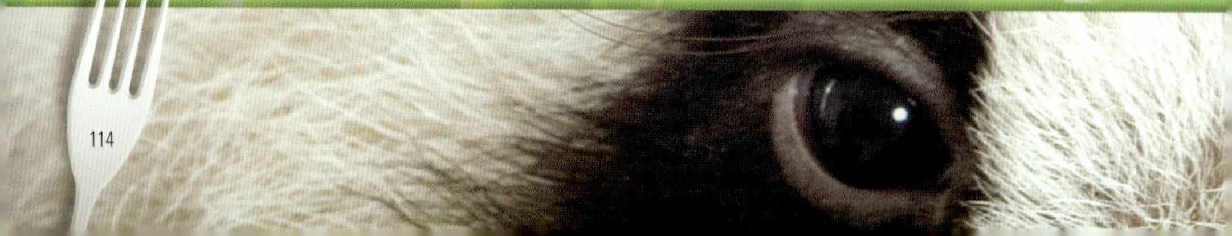

Interestingly, even though the Giant Panda is mostly vegetarian, the easiest way to catch it is with goat meat. It likes a little goat meat. And the Giant Panda's DNA can make all the enzymes needed to digest meat.

But if you try to feed the Giant Panda high-energy foods, and steer it away from bamboo, it suffers abdominal problems and becomes less fertile. And speaking of fertility, Giant Pandas breed very slowly.

The Giant Panda is a mystery. But one day, if we understand the gut of the Giant Panda, we might be able to help the species survive and thrive.

Saccharin Rots Rats' Regulation

Our human bodies have evolved in a world where sweet-tasting foods are more calorie dense. The body deals with the expected caloric load of sweet foods by various reflexes that control how much you eat, how the energy is used, and so on. But artificial sweeteners, such as saccharin, break this link between taste and energy – they give sweet tastes without the increase in calories.

How do we react to this trickery? Not very well, it appears.

A study by Susan Swithers and Terry Davidson of Purdue University in Indiana looked at rats eating as much as they wanted of regular rat chow. But first they were fed either yoghurt sweetened with glucose or yoghurt sweetened with saccharin.

Surprisingly, rats given artificially sweetened yoghurt ate more overall and had increased weight and body fat. They also had less of the normal increase in temperature that happens after a meal.

In the USA, diet drinks make up about 30 per cent of $70 billion annual soft drink sales. The Calorie Control Council says that about 200 million Americans consume zero- or low-calorie food products, and that about half of these people consume four such products every day. Many Americans use artificial sweeteners as an excuse to eat high-energy foods.

In the rats, the artificial sweetener broke down the link between sweet tastes and calories, and the rats ate more and got fatter.

Oh rats!

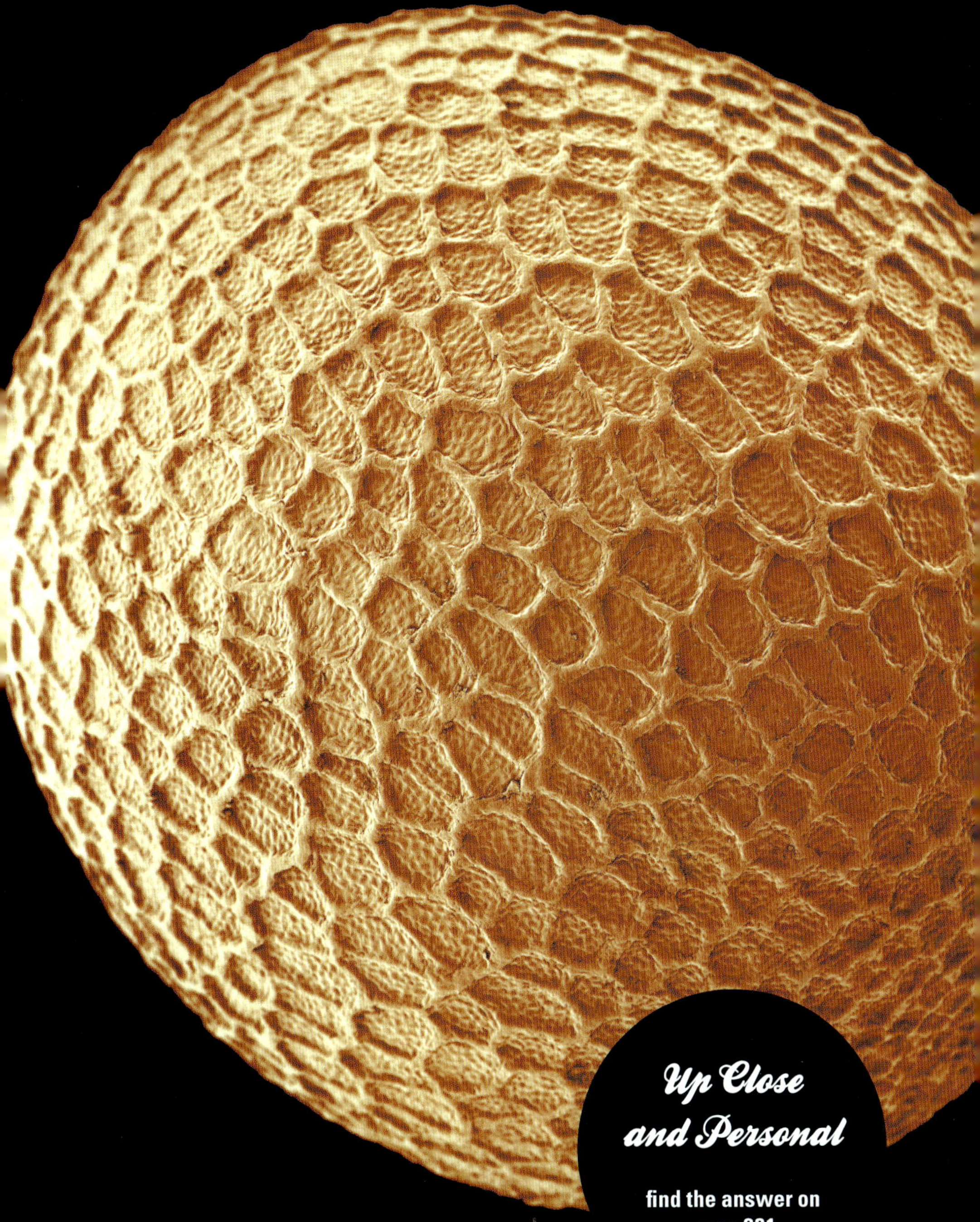

Up Close and Personal

find the answer on page 231

SHINKANSEN Series E3
SHINKANSEN Series E3

The Food Industry under the Microscope

Nutrition and diet cover an enormous amount of territory. Luckily, many people have written excellent and insightful books on aspects of this huge body of knowledge. I am particularly impressed by three authors.

US chef and author Michael Pollan has been writing about food, agriculture, drugs, gardens and architecture for over 20 years. His book *In Defense of Food: an Eater's Manifesto* does exactly what the title says. I love his definition of "Food".

David A Kessler was previously the controversial and dynamic Commissioner of the American FDA (Food and Drug Administration) who famously exposed the tobacco industry. He was also a paediatrician. His book *The End of Overeating* takes on the American Food Industry. He writes about how combinations of fat, sugar and salt change our brain chemistry so that we end up craving more of the same.

And Jennie Brand-Miller is Professor of Nutrition at the University of Sydney. (Personal disclosure here – we have worked together on various minor projects in the past, and I am a big fan of her work.) She is one of the world's foremost authorities on the Glycemic Index (GI) and on carbohydrates in general, and author of many books, including *The Low GI Diet*. It might surprise those on high-protein/low-carb diets but sugars were not invented by the Devil.

After reading their work, I realised their general wisdom and specific insights were too good to ignore. So I now offer three tiny summaries. But, of course, to get the whole experience, you should read their books cover to cover.

The Food Industry under the Microscope Part 1

In Defence of Food

Michael Pollan succinctly states: "Eat food. Not too much. Mostly plants." This is simple, but deep. Pollan writes from an American perspective, but there's crossover to most English-speaking Western countries.

Pollan defines Food as something your grandmother would recognise, and something that can spoil. Food should not be highly processed, nor should it contain a long list of unfamiliar, unpronounceable ingredients. It should not be full of fructose corn syrup. It should make no health claims. It should not be packaged. Ideally, you should be able to buy it from the person who has grown it. Like most chefs, Pollan advises you to cook with seasonal produce.

It might seem odd to have to define Food, but one of Pollan's key points is that a lot of what we eat *isn't* Food as he sees it.

Eating "not too much" makes perfect sense. Eat small portions, eat slowly, enjoy what you eat, and "listen" to your body for cues telling you you're full. Do not have seconds, and eat meals not snacks. It might sound obvious, but eat at a table. And try to eat in company, if possible.

Finally, eat "mostly plants". Yep, eat mostly plants (especially the leaves) remembering that seeds carry lots of energy.

More Food, Less Nutrition

Today's Food Industry generates more tonnes of food per hectare than ever before. But these increased yields come at the expense of quality.

Continuous growing of the same crops in the same soils removes the nutrients. In turn, these depleted soils lead to crops with fewer nutrients. Industrial farming tries to correct the soil depletion by fertilising with micro-nutrients. This is a short-term fix and doesn't work in the long term because the micro-ecosystems of healthy soils are ignored.

The USDA (United States Department of Agriculture) studied 43 crops that it consistently tracked since the 1950s. On average, the Vitamin C content in today's crops was down 20 per cent; iron was down 15 per cent; riboflavin was down 38 per cent and calcium was down 16 per cent.

As a specific example, an apple from 1940 contained *three times* as much iron as one of today's apples. These claims are consistent with UK figures. Pollan writes that "we are getting substantially less nutrition per calorie than we used to". Without any irony, a Food Industry magazine, *The Packer*, suggested "that this might be good for business, because people would now need to eat more produce to get the same nutritional benefit".

Modern fertilised crops grow more quickly. But they have less time to accumulate micro-nutrients from the soil, and have shallower root systems, which gives them less access to deeper, mineral-rich soils. The shallower root systems also mean that the crops cannot reach the deeper water tables – so they need more watering than slow-growing crops.

We also increase food yields by using pesticides. This certainly increases the quantity. But organic crops have to protect themselves from pests and disease so they make more phytochemicals. Many phytochemicals are antioxidants, which seem to be beneficial in our diet. While it might not sound practical to feed our world with organic food, it's worthwhile knowing what we're losing when we give up non-organic farming techniques.

Fatter and Sicker

Lots of people are trying to explain the current obesity epidemic in the USA. Pollan blames the way that food is processed, regarded and eaten. He thinks that American Protestant Puritanism, which traditionally scorns enjoyment of food (and pretty much anything else), is a factor. But perhaps more significant is that, nowadays, it's simply too easy to get food anywhere and at any time.

Reductionist vs Holistic

While eating may have become easier, it's also become unnecessarily complicated. For some people, eating is seen only within a reductionist and nutritional context. Pollan disagrees with this approach – to be balanced, he says, eating needs a cultural context as well.

The science of Biochemistry has given us an extra understanding of Food. But Biochemistry can be misused to give us a too-narrow view of Food, by looking only at its individual chemical constituents. "Reductionism" is the philosophy of examining in detail the individual parts – but sometimes missing the overall picture. So a Food Reductionist might think, "I should eat this product because it has Vitamin P*Q, Alpha-2 fats and lysine-free protein". The "Holistic" view would be, "I eat this apple because I'm hungry, and because it's delicious, and because I know its history".

A funny example of Reductionism comes from the American Seventh Day Adventist Dr Kellogg early in the 20th century. Dr Kellogg believed that animal protein was especially harmful to humans. It supposedly caused people to masturbate and toxic bacteria to flourish in the colon. This is why he invented Kellogg's Corn Flakes – which have outlasted his theories.

Pollan writes in great detail about "Nutritionism" – the ideology of labelling food nutrients as "good" or "bad". Unfortunately, "Nutritionism" opens the door to faddish dietary advice relating to cutting out proteins/carbs/fats/refined carbs/fibre, and so on . . .

Governments have a role in looking after their citizens – but Government Policy Makers are vulnerable to Food Industry lobbyists. It's very difficult for Policy Makers to recommend eating less of any single food item (for example, meat), because of pressure from Food Industry groups. Instead, Policy Makers come out with soft messages about food nutrients – for example, eat less saturated fat (which, by a coincidence, is mostly found in meat)!

A lot of the research (which ultimately advises us on what to eat) has been based on "whole food" intake rather than "nutrient" intake. So the advice about nutrient needs has only been extrapolated from this research.

And there's another problem: it's difficult to set up studies of very large numbers of people, where the scientists can totally control what the subjects eat over a long period of time. So some studies in nutritional science are based on rather rough data – where people report what they eat. In reality, most people underestimate what, and how much, they eat. They are also more likely to claim that they eat what they believe to be good food choices, rather than what they actually eat.

Problems with Reductionism

The low-fat campaign in the USA has coincided with increased rates of obesity, as presumably Americans did their best to eat more and more low-fat food!

The Food Industry loves advising us to eat certain nutrients, but this is often because they can package and sell them. Then they can try to make food "healthier" by adding or removing a particular nutrient. However, even in the early 21st century, nobody knows everything that a whole food contains.

A good example is margarine. Back in the 1950s, margarine was promoted as the "smarter" choice – after all, it contained "polyunsaturated" fats, as well as added vitamins. Some of these unsaturated fats were also Trans Fats. Unfortunately, at the time, we didn't know that Trans Fats were bad for your cardiovascular system. This led to the famous 2006 headline in *The New York Times*, "Killer Girl Scouts". It claimed that Girl Scouts killed more Americans than al-Qaeda – because the Girl Scouts raised money by selling cookies door to door, and these cookies were loaded with harmful Trans Fats.

The Food Industry Triumphs

The more processed the food, the more chances there are to modify it. It's pretty simple to add stuff to pureed apple, or apple juice – but it's a lot harder to alter a whole apple. Even so, the Food Industry is trying.

Generally speaking, there have been no changes in the nutritional quality of whole foods – only changes in how we perceive them. Back in the 1980s, everyone perceived avocadoes as "bad" because they were high in fat. But now we perceive them as "good" because they carry lots of monosaturated fats, which are (currently) labelled as "good".

When the Food Industry reduces food to nutritional components, they can market "designer" foods. Profits increase as the food is processed more and more.

Pollan says that the Food Industry also uses dietary guidelines to their own benefit. They find new reasons to process food, rather than dealing with the issue that the processing itself is the problem. The Food Industry makes the profits, but the Government has to pay for treating the chronic diseases that result from a poor diet.

Conclusions

According to Michael Pollan, we humans can healthily adapt to many different diets. The exception is the Western highly refined carb diet of flour and sugar. He quotes a study of Australian Aborigines who returned to a traditional bush diet for seven weeks. Even in this short time frame, the traditional diet led to both a decrease in weight and reversal of metabolic abnormalities such as Type 2 Diabetes.

In nature, the sugar "fructose" is found in ripe fruit. As such, it comes and goes with the seasons. In ripe fruit, fructose is combined with fibre and other valuable dietary component micro-nutrients.

But today fructose is added to processed foods and makes up about 20 per cent of the calories in the typical American diet. This provides quantity, not quality.

Pollan promotes the concept of "food synergy", where we see whole foods as more than just the sum of the nutrients. In other words, it's better for you to eat the whole grain rather than an equivalent amount of fibre and vitamins.

Unfortunately, it is harder than it might seem to eat unprocessed food. Industrialised food production is part of the chain. Perhaps the best piece of advice is to just "Cook and garden!"

Peppery Wine Perfume

Pepper has been around for millennia. When Alaric and the Visigoths besieged Rome, they demanded about 1400 kilograms of pepper as ransom. Pepper was an important spice on the ancient Spice Roads, and was a major reason for Columbus setting off on what he thought was the short way to India. Pepper originally came from the south-western coast of India, but today it's extensively cultivated in Brazil and South-east Asia.

Wine has also been around for millennia. But it was only as recently as 2008 that Australian scientists finally identified the compound "rotundone" that causes the pepper aroma of Shiraz and, of course, pepper.

The human nose is incredibly sensitive - it can detect rotundone at levels as low as 16 billionths of a gram/litre of red wine. Rotundone is found in black and white peppercorns at levels about a thousand times higher than in "peppery" wines. Surprisingly, about 20 per cent of people simply cannot detect this smell in Shiraz wines - even if the rotundone levels are increased 500 times.

It's good that we have finally found this chemical. After all, it would be hard to explain describing your wine as peppery if it didn't smell like peppercorns.

But I would be ever so thrilled if somebody could find me a reason to have wine described as smelling of cigar box!

Up Close
and Personal
find the answer on
page 231

Bacteria Burger

Hamburgers can kill. Or to put it more accurately – the bacteria that live inside hamburger meat can kill.

Ever since penicillin kicked off the modern antibiotic game about 60 years ago, bacteria have been evolving. Nowadays, diseases like gonorrhoea and leprosy, which were once sensitive to penicillin, are quite resistant.

Bacteria build up their resistance by growing and mutating in animals that carry low levels of antibiotics. The fact is, some bacteria are naturally more resistant and you need only one tough bacterium to start off the entire next generation. Give them a few hundred generations – which might take only a week or so – and the next thing you know, almost all the bacteria are immune to this low dose of antibiotic.

So what sort of animal runs around on low-dose antibiotics?

Usually not the human animal. When you go to the doctor's the doctor gives you a *big* dose of antibiotics that pretty much kills all the bacteria. Also, you are told to take the antibiotics to the very end of the course, even after you feel better. So there should be no bacteria left hanging around in you to mutate and multiply. (If you don't finish the course of antibiotics, you might allow some hardy, naturally resistant bacteria to survive and pass their resistance on to their descendants.)

But intensively farmed animals, such as poultry and beef cattle, walk around loaded with low-dose antibiotics. At one time, it was wrongly thought to be "cheaper" to give antibiotics to *all* of the beef cattle in the developed world, rather than wait until they actually got sick and then treat them. As a result, these antibiotic-fed cattle have a mini-war between bacteria and antibiotics going on inside them!

So the Wheel of Buddha rolls around – the animal is killed, and the raw meat goes to the hamburger shop. There it is cooked and the bacteria die.

But there's a catch: raw meat often ends up on the bread board. While it's sitting there, a few canny and clever bacteria jump off and wait. When the salad comes along the bacteria jump onto the salad. Now you've got salad with bacteria on it – and some of these bacteria are quite nasty and will not get cooked.

Antibiotics have been banned in European animal feed since 2006 (apart from two specific types that can be given to poultry). However, they are *not* banned from animal feed in the USA and in 2000 over 70 per cent of all antibiotics dispensed were given to animals we eat, such as cows, pigs and chickens. Most importantly, these antibiotics were given in the total absence of any disease.

We have sacrificed the long-term effectiveness of antibiotics, which are some of the most important drugs we ever devised, for cheap meat and, in retrospect, perhaps we made the wrong choice.

Clever bacteria which have developed resistance to antibiotics include *Samonella*, *E. coli* and *Staphlococcus aureus*. There are varieties of *Salmonella* that are resistant to nine different types of antibiotic. In the USA each year, *E. coli* causes around 73,000 illnesses and 61 deaths. And of the 47 per cent of meat and poultry contaminated with *Staphlococcus aureus*, over 52 per cent of these bacteria are resistant to at least three separate classes of antibiotics.

A really nasty type of *Staphlococcus aureus* – Methicillin-Resistant *Staphlococcus aureus* (or MRSA) – has evolved a new strain (ST398) that has found a happy home in modern pig farms. Between a quarter and one-third of all American pigs now carry that strain of MRSA. Each year, the disease kills more Americans (around 18,000) than AIDS.

So how do we keep farm productivity high without creating these monster diseases?

In Denmark, annual agricultural antibiotic use dropped from 210 tonnes to 9 tonnes between 1995 and 2000. Antibiotics were used only to treat sick animals. As a result, the incidence of antibiotic-resistant bacteria dropped enormously. Not only that, better animal husbandry kept the productivity virtually at its previous level. This is a Good Start.

Nevertheless, next time you're in a hamburger shop and see someone putting raw hamburger meat on the bread board, it might be a good idea to ask for a well-cooked salad, too.

Coffee Good Enough *Not* to Drink

Forget waking up to smell the roses. The mere act of smelling coffee is enough to wake you up – at least, if you're a rat.

A Japanese study compared well-rested rats with sleep-deprived rats. They then compared the activity of 11 genes essential to normal functioning of the rat brain – before and after they gave them the smell of coffee (not the liquid, just the smell). As they expected, 9 of the 11 genes were *not* working properly in the sleep-deprived rats. But giving these rats a whiff of coffee aroma was enough to restore these genes back to normal.

It turns out that there are more chemicals than just caffeine in coffee. "Quinic acid" imparts a slightly sour taste. "Niacin" (yep, Vitamin B3) is released when the heat acts on another chemical naturally present in coffee called "trigonelline". And there's "putrescine" – which smells like faeces, makes spoiled meat poisonous and comes from the bacterium *E. coli* attacking amino acids. We don't know which of the many, many chemicals naturally present in coffee wakes the rats up.

Even so, this research might give Bill Clinton a fresh reason to inhale – the coffee, that is!

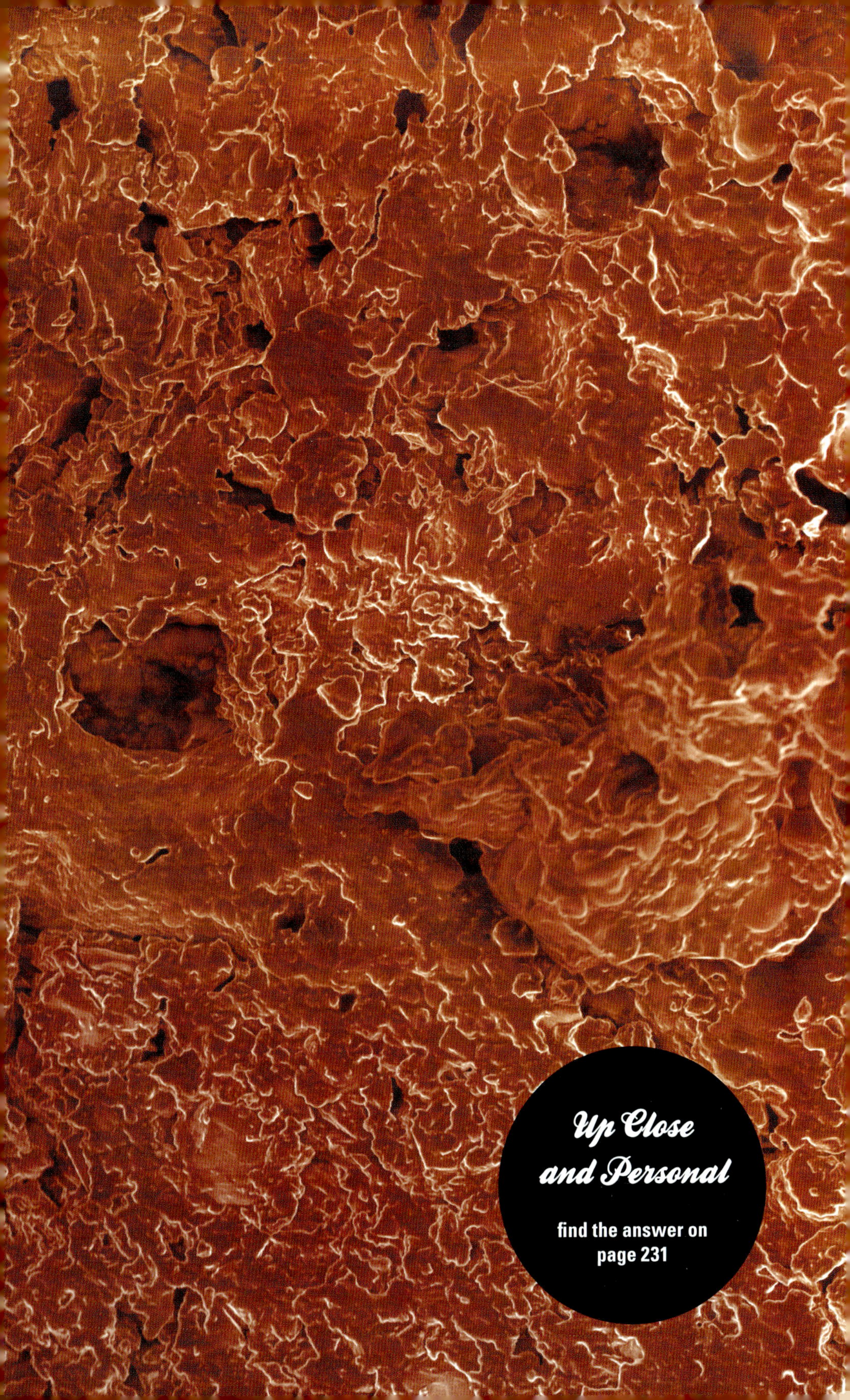

find the answer on
page 231

Onions and Tears

When you cut onions, why do your eyes leak salty water? (Okay, I'm talking tears.) And why does the same "leaky-thingie" happen when you're upset? And why are we humans the only animal to do this? And does it actually make you feel better?

Weeping 101

One of the funnest aspects of reading scientific literature is checking out The Definitions (really and truly). Back in 1993, Vikram Patel wrote:

> "Crying can best be defined as a complex secretomotor response that has as its most important characteristic the shedding of tears from the lacrimal apparatus, without any irritation of the ocular structures. It is often accompanied by alterations in the muscles involved in facial expression, vocalizations and in some cases sobbing – the convulsive inhaling and exhaling of air with spasms of the respiratory and truncal muscle groups."

So now you know . . .

Crying in public used to be considered normal – until the Industrial Revolution. The Hebrews, Greeks and Romans had special little bottles, called "lacrimatories", into which they would cry at funerals. These bottles would then be buried with the departed as a tribute and mark of respect.

Hippocrates thought that crying would safely release "excess humours" from the brain (where they were obviously causing some kind of problem). Aristotle wrote that crying "cleanses the mind" of suppressed emotions. The Roman poet, Ovid, believed, "It is a relief to weep; grief is satisfied and carried off by tears".

St Francis of Assisi apparently went blind from crying too much. One 1579 explanation for crying, with a rather poor understanding of physiology, claimed, "When the brain is compressed it ejects great quantities of tears". In Renaissance Europe, people suspected of being witches or werewolves were ordered to cry on demand. If they could not cry, they were deemed to be guilty, and then killed. Charles Darwin wrote, "Crying is a release for eyes over-engorged with blood by excessive emotion".

In modern society, crying often happens at the most sublime of events (births, marriages and deaths) as well as the most mundane (a squabble over nothing important).

It turns out that there are three kinds of tears that come from your eyeballs, and that there are three kinds of crying.

But, first, a recipe.

Long Slow Onion Cooking

Russ Parsons of *The Los Angeles Times* describes how to turn onions into a "deep mahogany reddish-brown with a marmalade-like consistency". Yum. These "caramelised" onions can be added to toasted bread (making *crostini*), used as a pizza topping or a flavouring base, or simply mixed through some cooked pasta. Add a bit of parmesan and you're away!

First of all, get *lots* of chopped brown onions, "put them in a pot with a little oil and some salt: cook and stir; cook and stir; cook and stir . . ." and so on.

You absolutely must *not* burn the onions, so you need the heaviest pot you can find to spread the heat evenly. Stir them at least every 15–20 minutes, decreasing the time between stirring down to 5–10 minutes in the crucial final stages.

0–20 min – covered pot, med–low heat.

45 min – lots of liquid sweated from the onions, remove lid, medium heat.

75 min – onions have reduced in volume, lowest possible heat, stir every 10 min.

120 min – hardly any moisture left, onions have darkened, stir every 5 min.

180 min – onions are a reddish-gold colour and have a lovely smell, stir every 5 min.

180–240 min – when onions are a little darker, are sizzling despite the low heat and are still moist and flexible, stop.

Eat them immediately, or seal and store them in the fridge.

Tears 101

Tears are mostly produced in the upper outer corner of the eye sockets. The drainage points are tiny pipes (0.3 millimetres in diameter) called puncta. They're located in the lower inner corner of the eye sockets. This arrangement means that our eyes get a very efficient Cross Flow Lubrication. Each punctum drains into the nose and down the back of your mouth – which is why you occasionally taste "something odd" after you have been crying.

Blinking helps spread the tears evenly over the eyeball, as well as sending any excess liquid into the puncta. Each blink lasts 0.3–0.4 seconds, and you blink every 2–10 seconds. Doing the maths, a person who blinks both long and frequently can, over a whole lifetime, spend a solid seven years blinking.

Reality is for People who Can't Handle Drugs . . .

You spend a lot of time each day *not* looking at the Reality Around You – and I'm not talking about Blinking.

Your eyes are constantly scanning in jerky movements called "saccades". During the middle of a saccade, you don't see anything and are effectively blind. Each saccade lasts up to 0.2 seconds, and you have about two of them per second.

This means that every day, you are *not* looking at Reality for about four hours – even though your eyes are open. That works out to about 12 years in an "average" 72-year lifespan. If "reality" is so important, how come we can ignore it so much, and not suffer any consequences?

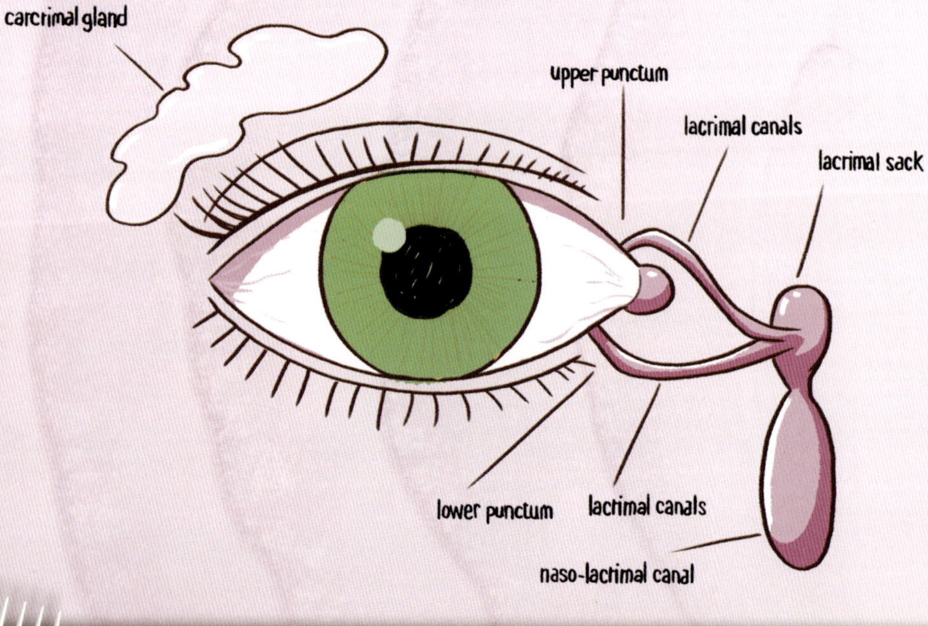

There are three different types of liquid in the "tear film" that cover your eyeballs. They're each made by different glands.

First, there is a Mucous Layer of tear film that rests immediately on your eyeball. It's made by Goblet Cells in the conjunctiva (a membrane covering the whites of the eyeballs and the inside of the eyelids). It sticks to and coats your eyeballs closely and evenly.

The next layer of tear film is called the Aqueous Layer. This liquid is made by the Lacrimal Gland in the upper outer corner of the eye socket. The Aqueous Layer is mostly water, with various proteins, antibiotics and minerals, and it protects your eyeball from bacteria, changes in temperature and saltiness, and so on.

The outermost layer is the Lipid Layer. This oily liquid is made by the Meibomian (or Tarsal) Glands, which are located at the edges of your eyelids. This layer coats the Aqueous Layer underneath, slowing down its evaporation – and stopping it from dribbling onto your cheeks.

Official State Vegetable

The state of Utah, in the USA, has the Spanish Sweet Onion as its Official State Vegetable. However, it has the Sugar Beet as its Official Historic State Vegetable.

Utah also has the Browning Pistol as its Official State Firearm.

Sweet Onions not so Sweet

You would think, quite reasonably from the name, that "Sweet Onions" are sweeter than regular onions – that is, that they contain more sugar.

Nope.

Sweet onions actually contain less sugar than other onions. They taste sweeter because they have fewer bitter, sulphurous chemicals.

This means that they are great for salads. But if you cook them the sulphurous chemicals evaporate, and they taste pretty bland.

Crying 101

There are three different types of Crying, and they are classified by their cause.

First, there's the crying that produces the regular Protective or Basal Tears. These tears moisten and protect your eyeballs as well as catch any microscopic dust and whisk it away.

Second, there are the so-called Reflex Tears. They gush forth in response to larger particles, gases and other irritants, including bright light. They also appear when you eat something really spicy.

Finally, there are the Psychic (or Emotional, Crying or Weeping) Tears. We believe they're unique to humans. They happen in response to either intensely sad or happy emotions – and tear-jerker movies designed especially to make you cry.

It is widely claimed that Emotional Tears contain higher than normal levels of three different chemicals. However, this research was done in the early 1980s with crude equipment, and recent studies with more sensitive equipment could not find evidence of these chemicals. For what it's worth, the three chemicals were claimed to be prolactin (a hormone associated with milk production), ACTH (a hormone related to stress) and leucine enkaphalin (a natural painkiller, similar to morphine in structure).

Emotional Crying

Women do Emotional Crying more often and for longer than men; babies more than children; and children more than adults – and we all do it more frequently both at night and in cold climates. Parents are very sensitive to the cries of toddlers, while non-parents are mostly oblivious. In Western society there are culturally approved events for crying, including births, weddings and funerals.

There are many theories behind Emotional Crying, but none of them provide all the answers. Two of the heavies in this field are Randolph Cornelius, Professor of Psychology at Vassar College in New York, and Ad Vingerhoets, Professor of Clinical Psychology at the University of Tilburg in the Netherlands. Over the decades they have examined some of the theories behind why our lacrimal glands are connected to the emotional centres in our brains.

There is the Getting Rid of Toxins Theory. The first problem with this theory is that it's a very inefficient way to move material out of the body – for example, the gut can remove far greater masses. The second problem is that most of the tears go via the puncta (at the lower inner corner of the eye sockets) straight into the nose and pharynx. They land on mucus membranes and are promptly re-absorbed.

Another Theory is that Crying is Communication. Certainly, babies and infants cry to communicate. Tears tell parents that something is wrong and prompt them to act. Adult crying shows others that we are vulnerable – our vision becomes blurred, and the deep emotions are difficult to fake (except, perhaps, for some actors and politicians). These intense emotions help bond us together.

Yet another Theory claims that Crying is Cathartic – that is, it helps us purge emotions and relieve emotional tension. This is a very popular belief, especially in the media. (Some 94 per cent of media reports, when they mention crying, glibly add the tag that Crying is Carthartic, and Good for You. However, the evidence for this is "mixed" – a scientific way of saying "not proven".) It seems that catharsis is more common if there is:

- social support during the crying (for example, somebody is giving you a hug)
- if there is only one other person present
- if the crying is done in situations in which the feelings of shame about crying are relatively low.

In many cases, crying leaves the crier more susceptible to mood disturbances – that is, worse off.

Onions are The One

Onions were found in the eye sockets of the mummy of Ramses IV. The Ancient Egyptians believed that they symbolised the Universe and Eternal Life, thanks to their spherical shape and repeating concentric rings.

The name "onion" probably comes from the Latin "*unus*", meaning "one".

Onion 101

We have been cultivating onions for around 7000 years. They have been credited with everything from making hair grow on bald heads, to giving valour to Alexander the Great's troops – perhaps not very reliably.

More realistically and more usefully, onions appear in the recipes and cuisines of practically every culture in the world.

You get a hint as to the answer to the onion–tears question by asking another question: why doesn't a whole onion have such a potent smell, or tear-making effect, *before* it's cut?

Why Make Irritating Chemicals?

Animals can flee danger but plants cannot. So onions have evolved irritating chemicals to keep hungry herbivores away.

Mind you, a fungus has turned the tables on the onion, and has specifically evolved to track down this irritating onion chemical. It then attacks the onion, completely unfazed by the irritating chemical.

When an onion is cut (or crushed), the onion cells are damaged and a three-step chemical process occurs.

First, the cells release an enzyme called "allinase". (By the way, onions have unusually large cells which are easily seen under a cheap microscope. This makes it more likely for the knife to crush or bruise them, rather than slip in between them.)

Second, the allinase attacks a chemical called "1-propenyl-L-cysteine sulphoxide" and turns it into another chemical, "1-propenylsulphenic acid". (By the way, don't worry too much about the exact names of the chemicals. They do have other names as well, but I'm using the ones given in the peer-reviewed science journal *Nature*.)

Third, this chemical is then turned into the irritant gas "propanthial S-oxide" – also known as Lacrimatory Factor. This is the irritating chemical that makes you cry. It leaves the damaged onion and floats through the air.

When this chemical lands on the conjunctiva (the membrane covering the eye), the nerve endings become irritated. They send signals to the brain that tell the glands of each eye to make tears, which then wash the nasty chemical away.

Onions Without Tears – 1

Back in 2002, Japanese scientists made a surprise discovery about the biochemistry of the onion. The chemical pathway that makes the irritating, tear-inducing Lacrimatory Factor is different from the pathway that makes the onion's flavour chemicals. There is a little crossover, but they are mostly very separate. The scientists wrote that it might be possible to genetically modify an onion so that it did not make the tear-inducing Lacrimatory Factor, but was otherwise a normal onion.

In 2008, the New Zealand-based Research Institute for Crop and Food announced that they had managed to switch off the tear-inducing gene in onions. They also claimed their new onion would retain all the typical flavour. But the scientist involved, Dr Colin Eady, said it would be 10 to 15 years before the Stingless Onion was commercially available.

Onions Without Tears – 2

For those of us with short attention spans who cannot wait 15 years for the stingless onions, here are some other options:

First, when cutting onions, always use a sharp knife. The cells in onions are uncommonly large. A sharp knife will cut through them cleanly and "bruise" them less – which means less irritant chemical in the air.

Second, cut them under water. This will soak up the irritant chemical. But cutting under water might make you more clumsy, and more likely to cut yourself.

Third, chill the onions first in the fridge. Less of the irritant chemical is released at the lower temperature.

Fourth, use a fan to blow the irritant chemical away from you.

Fifth, wear goggles and a snorkel. When I was a penniless medical student living in a squat I used to suffer terrible watering eyes whenever I cut onions. Swimming goggles reduced the tears a lot, but not totally. I then noticed that if I held my breath so that no chemicals entered my nose, my eyes did not water. The obvious next step was to wear a snorkel. *Voilà*, absolutely no tears at all!

Crunching Chips

Eating is essential for survival. So it's not surprising that our body will "extend" the sense of "taste" by linking it to the sense of "smell". Sounds reasonable – after all, good food usually smells good. But what about linking "taste" to "sound"?

Psychologists at Oxford University ran a study in which their subjects ate chips while sitting in front of a microphone wearing sound-blocking headphones. The only way the volunteers could hear the sound of the crunch was via the loudspeakers inside headphones – which the psychologists controlled.

If the crunching noise was played back at a normal level, the chip eaters thought the chip was normally crisp. If the sound was made louder, or the high frequencies increased, they rated it crispier. And when the crunch or the high frequencies were muted, the chip seemed less crisp.

The crunching sound is made when your teeth break apart the microscopic fried, dry and brittle cells inside the potato. More cell fractures mean more volume, and proportionally more high frequencies. But stale chips are less brittle and soggier, as they have absorbed moisture – they sound quieter, with proportionally less high frequencies.

Heston Blumenthal, the British Molecular Chef from the Fat Duck Restaurant, was inspired by research like this. He designed a seafood dish that was served with individual iPods. Diners could listen to the "sounds of the sea" and enhance the experience of tasting the food.

The chef notes that the background noise level on planes is one of the factors that adversely affects the taste of food while flying. Only one of them, though . . .

Up Close and Personal

find the answer on page 231

Teaspoons Vanish

Medical researchers from the McFarlane Burnett Institute for Public Health in Melbourne could never find a teaspoon to stir their hot drink of choice. Since they were Scientists and not Lawyers, instead of writing a Strongly Worded Letter, they decided to study the rapid and constant disappearance of teaspoons from kitchens.

For their high-brow (high tea?) research they purchased, numbered and placed in kitchens 54 stainless steel spoons and 16 much nicer spoons.

After five months, 80 per cent of the teaspoons had vanished. The niceness of the teaspoon made no difference to how soon it disappeared. But teaspoons did vanish more slowly from smaller kitchens belonging to a specific group than from larger communal kitchens.

The scientists extrapolated their results to the greater population of Melbourne. They estimated that 18 million teaspoons per year go missing in that city alone. Those lost teaspoons would weigh as much as four blue whales.

They found a strange lack of previous research into Teaspoon Related Loss (TRL), so the researchers had to make their own conclusions about where all the teaspoons go. They were undecided as to whether to believe the teaspoons ended up living happily on their own Planet of Spoons, or to head down a more paranoid pathway and imagine the spoons were being deliberately antagonistic and disappearing on purpose just to show that humans did not control them!

Ultimately, all their research really proved is that you'll never satisfy the teaspoon needs of the nation given the rate at which teaspoons disappear. Perhaps all of the teaspoons have "shacked up" with all of the single socks.

The Food Industry under the Microscope Part 2

The End of Overeating

American doctor, lawyer and author David Kessler's book *The End of Overeating* is astonishing in its insights. Among other things, Dr Kessler points out the massive contradictions about food that exist in the USA (and in many other English-speaking Western countries).

Like Michael Pollan, Dr Kessler is interested in how the Food Industry manipulates food. Specifically, how it has modified food to make it *addictive*.

The human body evolved so that we eat when we are hungry and stop when we are full. But the goal of the Food Industry (Big Food) is the exact opposite. They manufacture a product that *stimulates* your appetite, so you eat more of the product. They do this by using a quirk of the human brain.

Let me point out that by the "Food Industry", I do not mean the market gardener, local providore or local food-growers' cooperative. I'm talking about the multi-billion-dollar multinational companies who sell processed foods that are far removed from their whole food origins.

Success for the Food Industry

For most of our many-million-year evolution, food was hard to come by. Sugars, fat and salt were rare and precious, and because we didn't get to eat them very often, our brains became hard-wired to enjoy them. They improved our chances of survival.

But in the early 1980s, the American Food Industry realised how to capitalise on this. They began to tailor-make products that combined sugar, fat and salt with a gorgeous "mouth feel". The effects showed up in two different ways.

First, sales of highly processed foods rose magnificently, and so did company profits.

The second effect was recognised by Dr Katherine Flegal, a Senior Research Scientist at the US Centers for Disease Control and Prevention.

While analysing data gathered from a US Federal Government survey into health and nutrition, Dr Flegal noticed an unusual change in the pattern of people's weight gain in recent decades when compared to the preceding hundred or so years.

Previously, American adults would normally gain a few kilograms between the ages of 20 and 40, but lose them in their sixties and seventies.

Now the configuration was completely different. Some 20 million Americans, or 8 per cent of the population, had quite suddenly (over a decade or so) become overweight.

Making Hyper-palatable Food

How had this happened?

Dr Kessler realised that many Americans were constantly fighting the desire to *overeat*. The *Journal of Clinical Investigation* reported that "Kessler theorizes that after being exposed to hyperstimulating foods, some individuals develop what is known as conditioned hypereating".

"Hypereating" sounds bad and "conditioned hypereating" sounds even worse. It basically means that Americans had switched from eating because they were hungry to eating because their appetite was permanently aroused.

The Food Industry quickly and cleverly learned that manufacturing "hyper-palatable" products (I hesitate to call them a Food anymore) was a simple, two-stage process.

First, you incorporate fat, sugar and salt into everything you make.

Second, you Load the fat, sugar and salt into the core ingredients, or you Layer them on top or underneath. Or you do both.

For example, say your core ingredient is chicken. The poultry factory deep-fries the chicken – so the fat is Loaded into the meat – and then freezes it for transport. The restaurant then fries the chicken again, which increases the Loaded fat. So you have fat on fat. Then you serve the chicken with a sweet and salty dipping sauce – that's called Layering.

If your chicken were a gun, it would be Locked and Loaded. But it's not. It's Loaded and Layered (with sugar on salt on fat on fat). But it's almost as dangerous over the long term.

Say you have a potato as your core ingredient. It's a carbohydrate – a bunch of sugars joined in a chain. Cut the potato into chips, deep-fry them, and fat is Loaded through the substantial surface area. (The thinner you cut the chips, the bigger the surface area, and the more fat they can carry.) Layer the chips with cheese, sprinkle lots of salt on them and feed them to the consumer. You have salt on fat on fat on sugar.

Of course, Cheesy Chips are yummy – which is good for the manufacturer – but they're very bad for you.

We Love Hyper-palatable Food

Unfortunately, most hyper-palatable products don't fill you up. You just keep on eating and eating until you've eaten a ridiculous amount. No wonder the manufacturers love them.

But how do they make the consumer love them?

Taste, for one.

The consumer won't love the product if it has too much or too little sweetness, saltiness or fattiness. There is a "bliss point" for these three ingredients where they each taste divine and it took the Food Chemists only a few years to find it. They then mixed in flavour enhancers in all kinds of combinations to "improve" the taste even further.

Another important factor is "mouth feel".

This is why the Food Industry loves Trans Fats – they are ridiculously easy to manipulate to give the product any texture you want. You can have solid, crunchy, flaky, gently flowing or any combination of the above. Trans Fats also act as a lubricant – so the product slides down into your gut more easily. In the Good Olde Days, people used to chew their food 25 times before swallowing it. These days it's down to 10 chews. Less chewing means you can eat faster and stuff in more calories.

Shame about the damage Trans Fats cause to the cardiovascular system. They are slowly being banned in developed countries. But the Food Industry only very rarely removes Trans Fats voluntarily.

Hyper-palatable Beats the Balance

Hyper-palatable food products can also alter your brain chemistry. They act as rewards even if you're not hungry – you still keep on eating after you're full. In fact, they are almost as "rewarding" as the drug cocaine. If you are dealing with stuff almost as "rewarding" as cocaine, you are dealing with an Addictive Substance. The product gets your attention, it stays in your memory, it changes your mood and it becomes your focus. You begin to spend your time thinking about your next meal.

Your body has a mechanism that balances the energy coming in with the energy going out. This keeps your weight reasonably constant. So if you have a big lunch, you would normally balance this by having a small dinner – but not if you're eating hyper-palatable food products.

Dr Kessler says that eating this stuff all the time effectively rewires your brain. Your brain becomes more sensitive, to the point of thinking constantly about the more rewarding hyper-palatable food products.

This is why Food Industry advertising uses emotional (rather than nutritional) appeals.

Hyper-palatable food products are linked to emotional pleasure. The emotional reward motivates you to repeat the action, over and over (the action of putting the product in your mouth). Eventually, the action becomes a habit and results in automatic eating – in other words, eating *without* enjoyment.

Our body's "energy in" versus "energy out" mechanism is ill equipped to fight this cleverly planned attack. It crumbles in the face of the firing of your brain's reward centres.

And so you eat another hyper-palatable food product.

If you try to not eat a particular product, you feel deprived – so, perversely, this increases the reward value of the food.

And so you eat another hyper-palatable food product.

You get some fleeting pleasure, and then you feel more out of control. The cycle repeats.

Ultimately, the Food Industry gets richer while you get fatter and sicker.

How to Run Free (as the Wind Blows)

Dr Kessler believes you need rules to break free of the compulsive eating loop. Rules get you back in control.

They're easier to stick with if they're simple. For example, "I don't eat chips or dessert".

Rules take away the need to exert willpower. As you follow the rules, you learn new responses – and you soon develop new automatic responses.

So get educated. Learn how to read food labels, and understand what's in the food (including Loaded and Layered ingredients). Most English-speaking Western countries no longer know the difference between a snack and a meal. When you have a snack, you should eat less at the next meal.

Maybe live in France for a year. In France, food is eaten only at mealtimes (and certainly never while walking on the street). People don't snack, and students never bring food into a classroom.

Eat smaller portions.

Be alert to situations that stress you into eating, so you can sidestep them. Distract your attention from food. Avoid yes/no debates in your head – concentrate on other goals.

Plan other activities instead of eating – talking on the phone, going for a walk . . .

And remember: exercise enhances wellbeing.

Chocolate Not Cheery

If chocolate can do even half the things that are claimed, surely it must be a Miracle Food. It is said to be both a stimulant and a relaxant (isn't that a contradiction?), an aphrodisiac and an antidepressant. But what if chocolate doesn't, in fact, cheer you up?

Research has shown that increased levels of serotonin in the brain can lead to an elevation in mood – in other words, the more serotonin you produce, the happier you feel. Professor Gordon Parker and colleagues from The Black Dog Institute in Sydney did some research into exactly how chocolate affects your mood. Could eating it increase your serotonin levels?

The scientists concluded that there are three major problems with this theory.

First, your mood can change *without* any change in your serotonin levels.

Second, your serotonin levels increase only if the protein content of the food you're eating is less than 2 per cent. (Chocolate contains around 5 per cent protein.)

Third, when you eat chocolate emotionally – as "Comfort Food" – it is more likely to prolong your depression, instead of bringing it to an end!

So the joy you get from chocolate is short and sweet – the pleasure of anticipation, the taste and the "mouth feel". Isn't that enough?

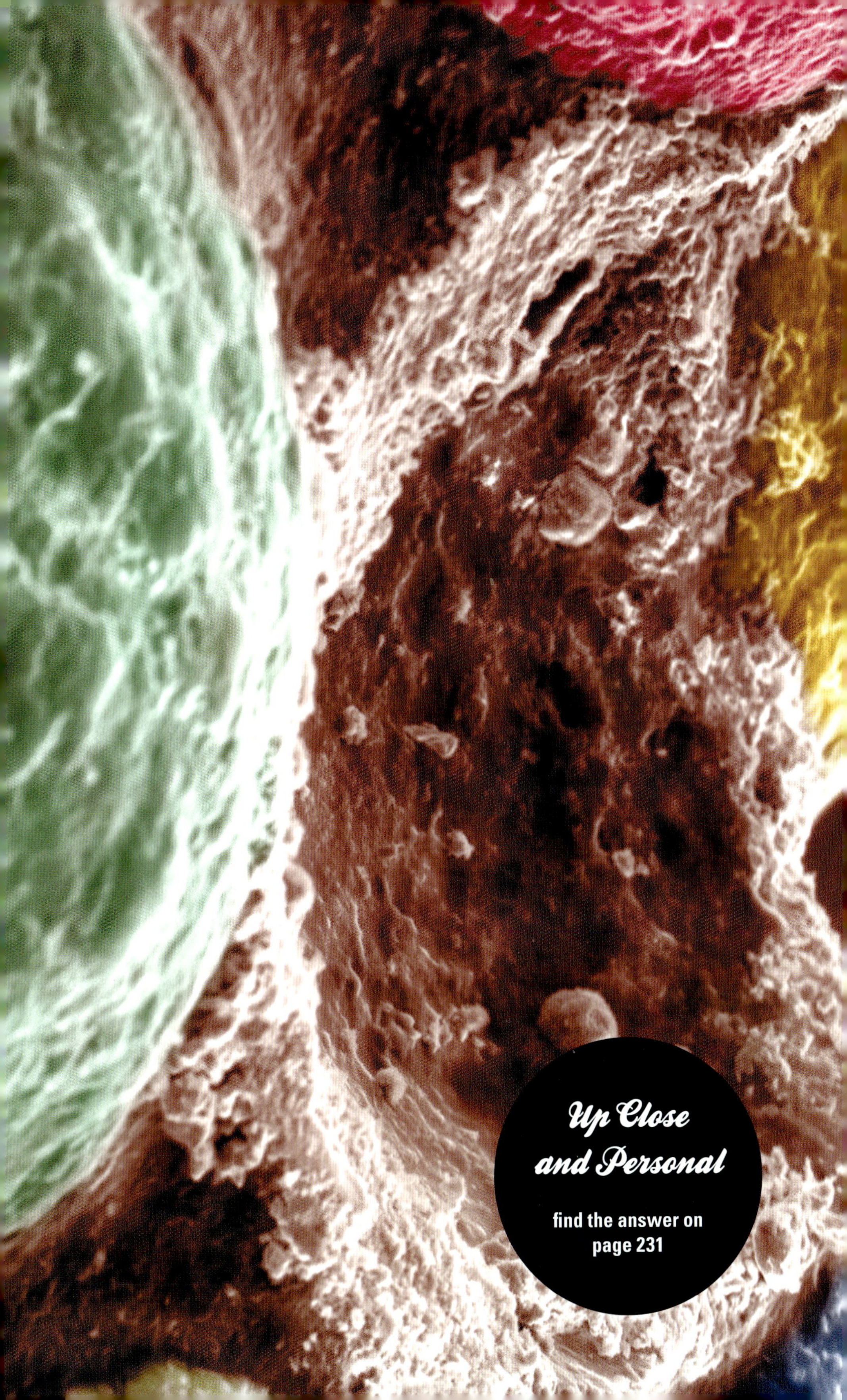

Up Close and Personal

find the answer on page 231

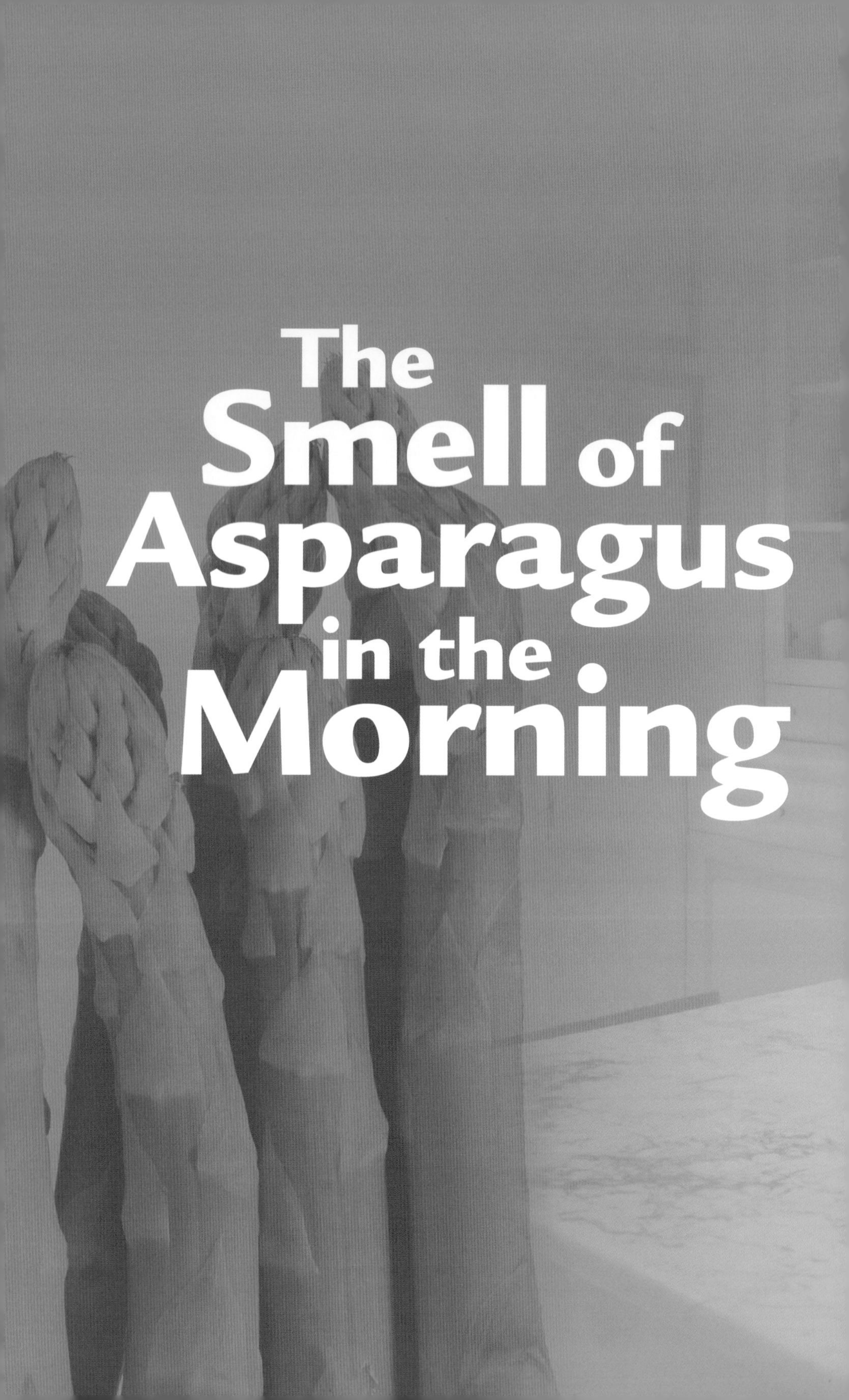

The Smell of Asparagus in the Morning

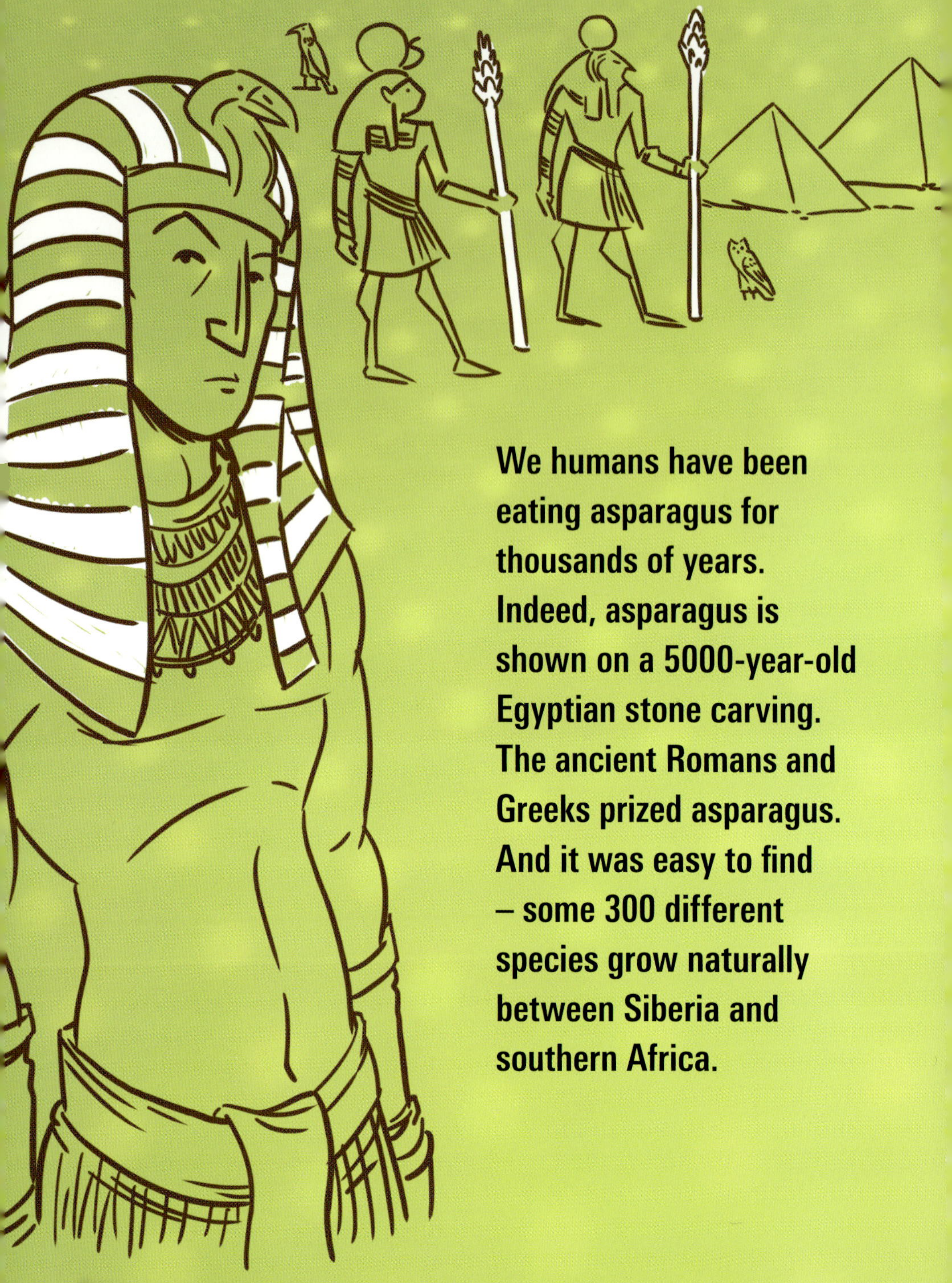

We humans have been eating asparagus for thousands of years. Indeed, asparagus is shown on a 5000-year-old Egyptian stone carving. The ancient Romans and Greeks prized asparagus. And it was easy to find – some 300 different species grow naturally between Siberia and southern Africa.

It's quite common that when we eat asparagus, shortly afterwards our urine smells very stinky – something like rotten or boiled cabbage, or even ammonia. But only some of us can generate, or make, this odour. Now here's something similarly surprising – only some of us can detect, or smell, this odour.

Because we humans are very human, everyone who can generate and detect this odour assumes that everybody else can. After all, they get proof about half an hour after eating some asparagus, when they go to the toilet. And this proof happens every single time.

And on the other side of the coin, those people who cannot detect or perceive this odour have no idea that asparagus can make your urine smelly – until someone tells them. Even then, they often find it hard to believe.

And today, smelly chemicals, like the ones in urine after eating asparagus, make it safer for us to heat and cook with natural gas.

Stinky Smell Saves Lives

To see the link between asparagus and natural gas, let's start with the skunk.

The odour of skunk is related to the asparagus-odour of urine. They ("skunk spray" and "asparagus urine") seem to be both caused by sulphur-containing chemicals called "thiols".

Many thiols have a repulsive odour. The human nose can detect the thiols in skunk spray at levels as low as 10 parts per billion (that's equivalent to 10 seconds in 32 years). But some thiols smell fine. For example, grapefruit thiol is quite pleasant at low concentrations, giving you a classic grapefruit smell. But at high concentrations it is unpleasant.

Thiols are deliberately added to natural gas, which has no odour, to warn you of its presence.

This routine was started after a terrible natural gas explosion in 1937, in the town of New London in Texas. The local school was not heated by the more conventional method of a large central boiler that sent hot steam to heaters in each room. Instead, it was heated by 72 gas heaters scattered throughout the school. That part of Texas was rich in oil, and at that time the school could get the natural gas for free.

Unfortunately there was a leak, and the natural gas accumulated in a closed crawl space that ran under the building's entire 77-metre length. Because natural gas is both odourless and colourless, there was no warning that it was present.

Shortly after 3 p.m. on 18 March 1937, an instructor switched on an electric sander. It gave off a spark that ignited the mixture of natural gas and air.

The explosion was heard many kilometres away. The roof was lifted entirely off and then crashed down again, collapsing the main wing of the school.

A block of concrete weighing over a tonne was tossed like a pebble across the parking lot and crushed a 1936 Chevrolet. Over 300 teachers and students died. It still remains the worst school disaster in American history.

Within a few weeks the Texas Legislature mandated that smelly thiols should be added to natural gas, making it easily detectable. This practice quickly spread across the USA, and then the whole world.

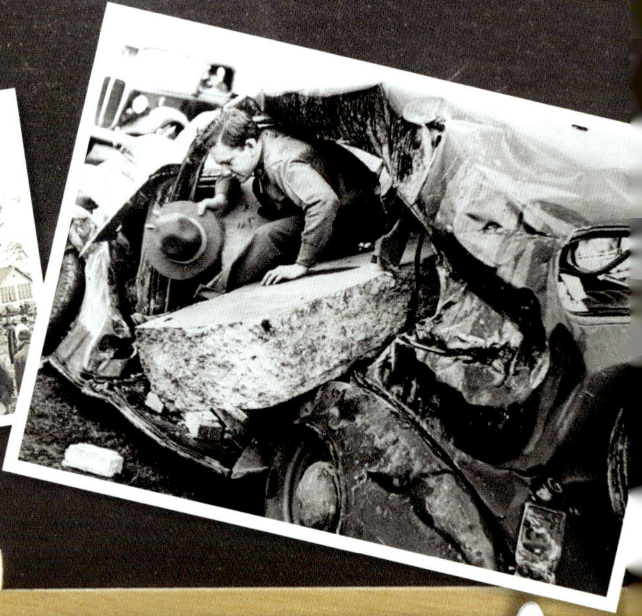

Very Stinky

Methanethiol, which is thought to be one of the smelly asparagus chemicals in urine, can be amazingly smelly.

In 2004, in Italy, an empty (that's right, empty, as in "exhausted" or "already used") methanethiol canister was being returned for refilling. This happens safely around the world, hundreds of thousands of times each year.

Very unusually, this particular empty canister sprung a leak when it was just north of Milan. Unfortunately, the winds carried the methanethiol stench across the eastern half of Milan. Thousands of people, some as far as 12 kilometres away from the empty (let me emphasise EMPTY) canister, phoned the emergency services number, convinced that there was an actual gas leak in the street outside their front door.

Asparagus Urine Smell – The History

Let's get back to basics.

Asparagus belongs to the lily family. Not all species are edible – in fact, some are poisonous. It's mostly grown in Peru, China and Mexico.

Asparagus was mentioned way back around 450–300 BC by various Greek writers such as Antiphon, Theopompus and Theophrastes. A century or so later, the Romans, a bit further along the Mediterranean, were perfectly familiar with asparagus. Around 200 BC, Cato the Elder gave excellent instructions on how to cultivate asparagus in his book, *On Agriculture (De Agri Cultura).* The first English author to write of asparagus was John Gerard, who in 1597 used the Latin word "*asparagi*", meaning "the first sprig or sprout of every plant, especially when it be tender".

Surprisingly, none of the ancient Roman or Greek writers ever mentioned the overpowering smell in the urine. The "offending" smell was first described only 2000 years later.

One of the earliest mentions was in the 18th century, by a physician to the French royal family, in his Treatise about Foods. He wrote that asparagus "eaten to Excess . . . causes filthy and disagreeable smell in the Urine". The Scotsman John Arbuthnot, who was both a mathematician as well as physician to Queen Anne, discussed this smell in his book on food published in 1731. He wrote, "Asparagus . . . affects the urine with a foetid smell (especially if cut when they are white) and therefore have been suspected by some physicians as not friendly to the kidneys; when they are older, and begin to ramify, they lose this quality; but then they are not so agreeable."

"a few stems of asparagus eaten shall give our urine a disagreeable odour"
100

That great American all-rounder, Benjamin Franklin, wrote that, "a few stems of asparagus eaten shall give our urine a disagreeable odour".

However, the great French author Marcel Proust was much kinder when he wrote that asparagus, "as in a Shakespearean fairy story transforms my chamber pot into a flask of perfume".

How come there is absolutely no mention of the stinky wee associated with eating asparagus until the 1700s? After all, those Ancient Greeks and Romans were just as smart as we are today, and surely must have had a sense of smell just as acute as ours?

One theory is that the stinky wee smell is related to the extra fertilisers used on asparagus in the 1700s. After all, it was only then that sulphur, sulphate and organic sulphurs were first used to perk up the flavour of asparagus. Here's some weak evidence for this theory.

If you grow onions and garlic in a soil that is low in sulphur, they will have a weak flavour and will lose their ability to make you cry when you cut them. So adding sulphur to some crops can have quite a dramatic effect on how pungent or irritating they are.

But another theory is that the farmers selectively bred the asparagus for some other quality (such as hardiness, or ability to resist extremes of weather), and bad urine odour accidentally came along with the extra hardiness.

Or maybe the Ancient writers were simply too polite to comment upon smelly urine.

Perhaps we'll never know . . .

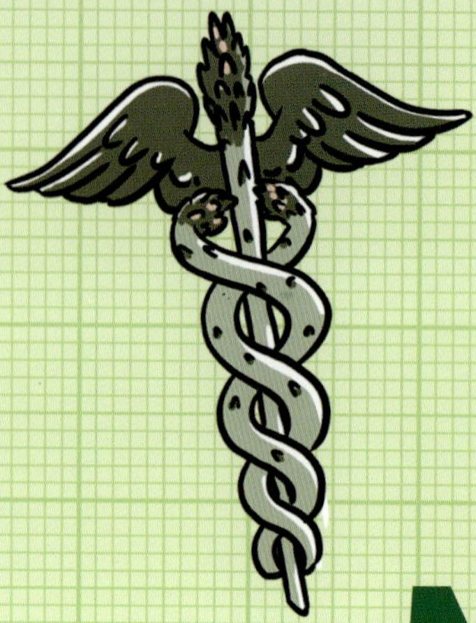

Medical Asparagus

Asparagus has long been used, with not a lot of success, to treat various medical conditions. These include cramps, toothache, constipation and even eye conditions. This lack of success is typical of most medications from The Bad Olde Days. On the other hand, asparagus is delicious and, like most foods, has various micronutrients, so we should eat it.

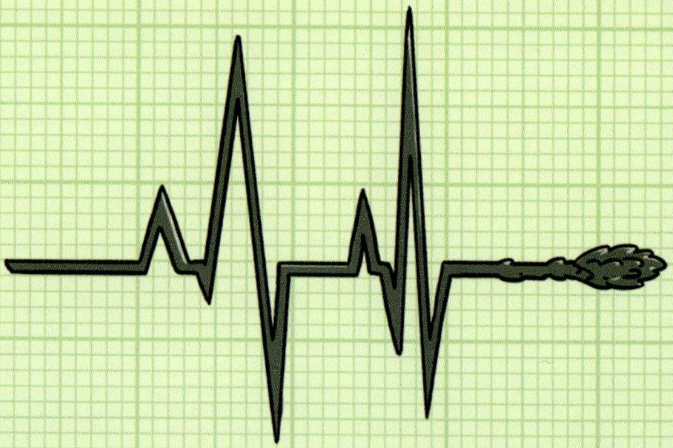

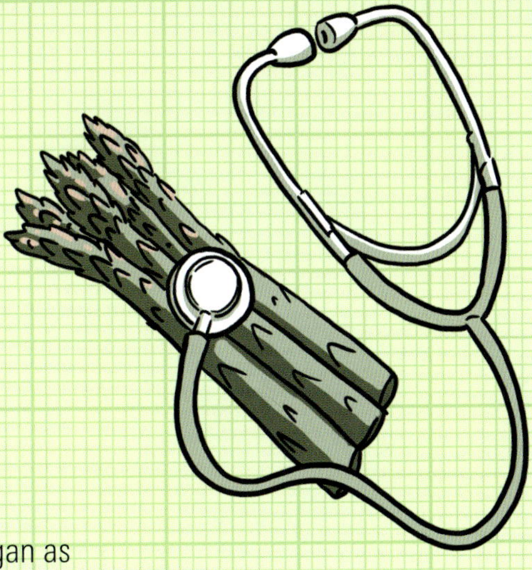

Normally, most people can eat asparagus with no troubles, apart from the stinky wee.

Very rarely, asparagus has been implicated in allergies. The first reports date back to 1880. They involved a Contact Dermatitis in asparagus pickers and in chefs who prepared and cooked it. It usually began as a Fingertip Dermatitis that could spread to cover the whole hand.

There is also a Contact Urticaria, which can show itself via swollen lips after eating asparagus.

Asparagus can also cause conjunctivitis (itchy eye(s), usually red), rhinitis (runny nose, sneezing) and asthma.

But the vast majority of people enjoy asparagus with no side effects – apart from the smelly urine.

There are a very few cases where the asparagus was not physically touched in any way, so perhaps the Nasty Allergic Chemical can travel through the air.

If you happen to be unlucky enough to be susceptible to asparagus, the offending chemical that attacks you is 1,2,3-trithiane-5-carboxylic acid. This chemical slows down or inhibits plant growth, and is at its highest levels early in the growing phase of asparagus. This allergy is well known in Germany, especially when dealing with young asparagus shoots.

Only Some People . . .

Not everybody can generate or make smelly asparagus urine, and not everybody can detect or smell it. And just to make it more complicated, not all the generators can detect it and not all the detectors can generate.

Not everybody can generate or make smelly asparagus urine, and not everybody can detect or smell it. And just to make it more complicated, not all the generators can detect it and not all the detectors can generate.

But don't worry: the situation gets even more confusing. The ability to generate and/or detect the typical asparagus smell in urine varies enormously around the world, depending on the study.

One study found that an Israeli population of 307 people had an amazingly low level of the ability to smell the asparagus in urine – around 10 per cent. A different study found this ability present in 25 per cent of a Chinese population.

One English study of 800 people from Birmingham showed higher percentages – about half could generate, and also detect, these chemicals in their urine.

In Philadelphia, a 2010 Monell Center study of 38 people showed that 96 per cent could smell the odour, while 92 per cent could generate it.

Winners = French

But the French win. A study of 103 French people showed that they *all* generated smelly urine after eating asparagus.

How come there is such a wide variation – from 10 per cent to 100 per cent?

It could be caused by genuine genetic differences between racial groups. But it could also be because unreliable methods were used to measure how the odour is generated or produced, people eating differing amounts of asparagus, different processing of the asparagus before (and after) it gets to your kitchen, or any combination of the above – and a few other factors that I haven't thought of.

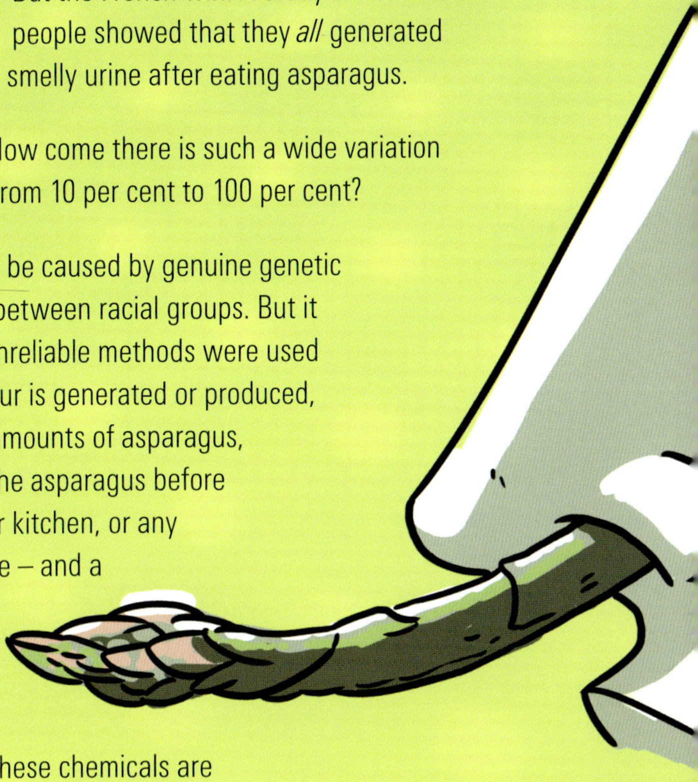

In addition, the actual techniques that isolate these chemicals are sometimes very lengthy and harsh and can damage other chemicals. In other words, perhaps the actual methods used in handling and testing the urine could have changed the chemicals originally present in the urine. These still unidentified chemicals could, in fact, be the real smelly chemicals.

Probably we'll have a much better idea why populations vary so much in a decade or two, especially with some help from Genetics.

Asparagus to Urine

The process begins with a chemical called Asparagusic Acid, which is naturally present in asparagus. The asparagus plant uses this acid to fight against attacks from aggressive critters such as parasitic nematodes. The asparagusic acid stops the nematodes from getting into the plant's tissues. This chemical has five atoms arranged into a ring with five sides – and two of the atoms are sulphur atoms.

You eat the asparagus, and then your gut breaks down the asparagusic acid into a whole family of chemicals rich in sulphur. These chemicals include methanthiol, dimethyl sulfide, dimethyl disulfide, bis(methylthio) methane, dimethyl sulfoxide, and dimethyl sulfone. These volatile chemicals are found in the space immediately above fresh urine – at least, the urine of some people about half an hour after they have eaten asparagus.

These same chemicals are also missing, or present only in very low quantities, in the urine of people who don't make smelly urine when they eat asparagus.

Chemistry tells us that the annoying smells seem to come from a whole bunch of chemicals rich in sulphur. But even today the scientists cannot fully agree exactly which are the odour-producing chemicals in Stinky Asparagus Wee . . .

White Asparagus

When my family and I walked 800 kilometres across Spain in 2009, we saw field after field of asparagus. At least, that was what we were told. We couldn't see the actual plants, because they were covered with thick layers of plastic – white on top to reflect the heat, and black underneath to block the sunlight. The lack of sunlight meant that the asparagus could not make chlorophyll. Chlorophyll is green, so the asparagus would grow up pale and colourless – white asparagus.

Back in the old days before black plastic, farmers would grow asparagus in caves, or under very dense foliage. In France, they give this "white asparagus" the name *"Argenteuil"*. This name comes from the region Argenteuil, where they grew the asparagus underground to make it white.

Preparing Asparagus

According to Marian Burros of *The New York Times,* "asparagus are like tomatoes and corn: they should be served only in season". She should know – she's written 13 cookbooks. So when Marian found a table piled high with freshly cut asparagus on a country road, she bought them all.

On the first and second nights she ate them freshly steamed. On the third night, "the asparagus were roasted in a very hot oven, drizzled with olive oil". Over successive nights, they were served as asparagus soup, "then risotto with asparagus and porcini; asparagus with a dressing of vinaigrette, chives and cornichons; asparagus, zucchini and parmigiano pancakes . . ."

Smelly Diseases

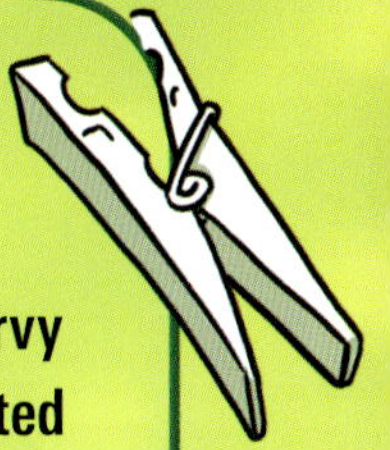

Patients with diphtheria smell "sweetish", while those with scurvy (vitamin C deficiency) smell "putrid". A diabetic coma is associated with a "fruity" smell, while yellow fever makes you smell like a "butcher shop".

The condition trimethylaminuria makes the holder of this inborn error of metabolism smell strongly of slightly old fish – hence the name "fish odour syndrome".

But generating or detecting the smell from asparagus in your urine has nothing to do with any disease state. There is some link with inheritance that we still don't understand fully – but there's nothing to worry about.

The Kilogram is Losing Weight

The "Kilogram" is losing weight and nobody knows why.

"Le Grand K", as it's also known, was cast to be the reference weight of The Standard Kilogram in 1879. It's a right-circular cylindrical block (39.17 millimetres in height and diameter) made of 90 per cent platinum and 10 per cent iridium. It lives in an underground vault in France and its official name is The International Prototype Kilogram.

The kilogram is the only unit still defined by a physical thing. For example, the "Metre" is no longer the distance between two marks on a metal bar. Instead, it's the distance travelled by light in a certain time.

Unfortunately, The Standard Kilogram is getting lighter – by the tiny amount of 50 micrograms – when you compare it to the average mass of 90-or-so other Official Standard Kilograms around the world. It's not much, but it is supposed to be the reference for the whole world. By the way, 50 micrograms is about the weight of a fingerprint! The range was between a loss of 665 micrograms, and a gain of 132 micrograms.

So scientists are looking at another possible internationally accepted Standard for the mass of the Kilogram – perhaps a certain number of atoms of carbon-12.

But it does seem unfair – if even the kilogram can lose weight, why can't we?

Up Close
and Personal
find the answer on
page 231

Before...

Vodka has No Calories?

When people talk about food, nutrition, vitamins, changing body weight and calories, they're dealing with a massive amount of information. One of the Big Issues, at least for some people, is how many calories there are in alcohol. To be specific, does vodka really have zero calories?

We can blame part of the confusion about this issue on a 1964 book that sold 2.5 million copies. *The Drinking Man's Diet* by Gardner Jameson and Elliott Williams enticingly told us that we could drink alcohol, reduce carbohydrates and lose weight – all at the same time! Beer went with nuts, wine cried out for cheese, and martinis, steak, bacon and eggs were the Staff of Life, according to the book's authors.

Then, the low-carb Atkins Diet in 1972 told us that "low carb = healthy". Following on the "low carb = healthy" theme, advertisements have been popping up telling us (presumably for our own dietary good) that some alcohols are "low carb" or "no carb" and are somehow better for us.

The "low carb = healthy" idea is consistent with the myth that if you want to enjoy a few drinks but not put on weight, you should drink vodka. After all, the myth claims that vodka has no carbs – and therefore it has no calories (or kilojoules). But while it is one of the most popular distilled alcoholic drinks in the world, and while the word "vodka" might come from the Russian word "*voda*", meaning "water", and while it *is* as clear as water, it's loaded with energy – more than carbs, but less than fat.

Calories or Kilojoules?

Back in 1824, the French physicist and chemist Nicolas Clément defined a new unit of energy. The "Calorie" was the amount of energy needed to raise the temperature of one kilogram of water by one degree Centigrade. It was given the symbol "C".

Confusingly, in 1929 the same word but with a lower-case "c" was defined as the amount of energy needed to raise the temperature of one gram of water (a thousand times smaller than a kilogram) by one degree Centigrade. This was given the symbol "c".

Physicists and nutritionists call them the small calorie or gram calorie (c), or the big calorie, large calorie, kilogram calorie, dietary calorie or food calorie (C).

When the new system of units took over, the new unit of energy was the Joule (J).

To keep it simple, one food calorie (C) is roughly equal to 4.2 kilojoules (kJ).

Even so, we still talk about burning off calories, achievements are called "milestones", and babies are still weighed in pounds, not kilograms . . .

Vodka 101

We humans have been fermenting and drinking alcohol for at least 7000 years. But fermentation can only achieve a maximum of about 15 per cent alcohol. Something else is needed. That something is distillation, and we've known about that for about 4000 years. Distillation involves gently heating the fermented liquid and boiling off the alcohol (which boils at a lower temperature than water) then catching the vapour and condensing it. Distillation can raise the alcohol content to 50 per cent or higher.

Vodka is around 40–50 per cent alcohol and seems to have come from Russia in the 14th century. What's known as the "Vodka Belt" covers Scandinavia, and Eastern and Central Europe. In Tsarist Russia, taxes of vodka sometimes accounted for a massive 40 per cent of the tax revenue.

The taste of vodka is often neutral, with nearly all the natural flavours of the raw materials (for example, potatoes) removed during distillation and filtration, and no extra flavours added. A typical glass of vodka therefore is roughly half water and half alcohol – and definitely contains no carbohydrates. This is quite different from beer, which *does* contain carbohydrates – and we all know that beer contains calories. So does this mean that because vodka has no carbohydrates, it has no calories?

To understand the answer, you need to know a little biochemistry.

Energy in Alcohol

In general, calories mostly come from three types of molecules in the food we eat – fats, proteins and carbohydrates. All three are made from atoms of carbon, hydrogen and oxygen, but proteins have additional nitrogen atoms. These different molecules share (from the point of view of calories) an important structural feature – they each have carbon and other atoms joined to each other by chemical bonds. There is significant potential energy stored in each of these bonds because it takes energy to force atoms to link up with each other. Your gut liberates this energy.

Even though fats, proteins and carbohydrates have similar molecular structures, they are different enough to release differing amounts of energy.

Proteins and carbohydrates give you 16 kilojoules (kJ) per gram, while fats are more than twice as energy-dense – they can give you 37 kJ per gram.

Now here's the rub – while alcohol is not included in the basic food groups, it still contains calories. In fact, with 29 kJ per gram, it is actually closer to fat than carbohydrates in terms of energy.

A "shot" of vodka (30 grams) contains 15 grams of water (zero kJ) and 15 grams of alcohol (15 x 29 = 435 kJ).

A glass of wine contains five grams of carbohydrates (5 x 16 = 80 kJ) and 13 grams of alcohol (13 x 29 = 377 kJ) giving you a total of 457 kJ.

Regular beer contains 13 grams of carbohydrates (13 x 16 = 208 kJ) and 14 grams of alcohol (14 x 29 = 406 kJ) giving you 614 kJ.

New "low-carb" beers are virtually identical in energy content to the old "lite" beers – they contain a total of about 400 kJ, including about 3 grams of carbs (3 x 16 = 48 kJ).

All alcoholic drinks have about 400–600 calories – and a lot more if you add sweeteners and creams.

On the other hand, if you have a few vodkas, you probably aren't worrying about the calories. Any thoughts about gaining weight are probably quickly displaced by your increased light-headedness.

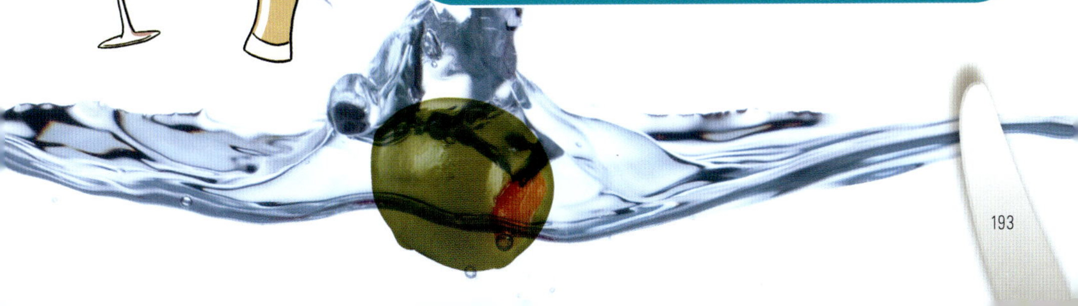

Ice-cream Headache

Why is it that some people get a headache when they eat ice-cream (or have a very cold drink, which might include a frozen margarita or a daiquiri) and others don't? It's no joke – "ice-cream headache" is a Real Condition. Some people also call it "Brain Freeze".

Ice-cream 101

Ice-cream is basically ice, fat and liquid. In most parts of the world where it's made, it has to contain at least 10 per cent milk fat. The delicious creaminess comes from the crazy way the air bubbles and microscopic ice crystals are held together by fat globules, with tiny pools of mushy sugary water in between. It is very difficult to get this just right, but if you can freeze your ice-cream mixture quickly enough, the bubbles will be tiny, giving part of that wonderful creamy "mouth feel".

Ice-cream has been around for a long time. In about 200 BC the Chinese were eating a frozen mixture of milk and rice. Two hundred years later, Roman chefs made ice-cream for Emperor Nero by mixing ice and snow (which the slaves had brought back from the frozen Apennine mountains) with nectar and honey. One of the first American mentions of ice-cream is from 1744, when a government official, William Black, wrote that the Governor of Maryland had served him the yummy stuff.

In 1850, Daniel Drake wrote a book called *Principal Diseases* of *North America*. In it, he described how ice-cream can cause an acute pain in the pharynx and a sense of coldness and sinking in the stomach – but he didn't say anything about a headache.

One of the first mentions in literature of an ice-cream headache is by James Jones (1921–77), who wrote a story called, wait for it . . . "The Ice-Cream Headache". In the story, a grandfather "loved to feed [the grandchildren] large doses of ice-cream on summer afternoons, would laugh at them gently when they got the terrible sharp headaches from eating much too fast, and then give them a general lecture on gluttony".

Signs and Symptoms

About 40–50 per cent of people will experience an ice-cream headache at least once during their lifetime. A Taiwanese study of 8359 junior high school students showed ice-cream headaches were more common in boys than in girls; that they increased with age, and that they were more likely to affect students who also suffered from migraines.

In your classic ice-cream headache a stabbing, aching pain begins a few seconds after you start eating and reaches its peak after about 30–60 seconds. The pain usually affects the centre line in the mid-frontal area of your head, but it can also affect just one side – near your temples, or the front of your head, or behind your eyes. Only rarely is an ice-cream headache felt at the back of your head. Happily, the pain begins to fade away 10–20 seconds after it starts – but it can sometimes last for up to five minutes.

There are many theories behind ice-cream headaches – but none of them fully explains what's going on.

One theory has it that they are caused by the action of one or more of the Cranial Nerves – either the Trigeminal or Glossopharyngeal nerve. They deal with sensation. They can pick up sensations from your pharynx, or your palate, or your tonsils. But they also pick up sensations from one of the membranes around your brain – the dura mater.

Now, you know when you put your hand into a bucket of ice-cold water, it begins to hurt. (By the way, if you say Bad Words while your hand is in the bucket, the pain will seem to be less. But this is not an excuse to use Bad Words for no reason.) In the same way, when you put something cold against your pharynx or palate or tonsils, you feel pain. Sometimes your brain interprets this pain as coming not from where the ice-cream is, but from the dura mater covering your brain – this is how you get the sensation of a headache from the pain in your pharynx, palate or tonsils.

This phenomenon is called "referred pain" because the pain is "referred" from where it actually happens (for example, the mouth or pharynx) to another location also connected to the same cranial nerves (the dura mater).

So you feel the pain while the Cold Stuff is in your mouth or pharynx – but not when it travels further down into your oesophagus or stomach, beyond the reach of these cranial nerves.

Another popular theory is that the pain is caused by sudden partial closing of the middle cerebral arteries (which deliver blood to the brain). One study (unfortunately involving only three people) measured blood flow in these arteries. In the two people who did get an ice-cream headache, the blood flow reduced by 20 per cent and 35 per cent. In the single person who did not get an ice-cream headache, the blood flow did not slow down.

There are many other theories, one involving sudden cooling of the blood entering the brain. None of them can explain why (in a very small number of people) the pain of the ice-cream is felt between the shoulder blades!

I guess this means that our understanding of headaches in general is pretty inadequate. But we do know one thing: if you want to reduce your chances of getting an ice-cream headache, eat more slowly!

Speed Eating

If you eat ice-cream quickly, you double your chances of getting an ice-cream headache.

The guinea pigs who tested this theory were 145 students from Dalewood Middle School in Hamilton, Canada. They were all given 100 millilitres of ice-cream. (By the way, every student who was asked to participate did so – nobody refused.) Half of the students ate all 100 millilitres of their ice-cream quickly – in less than 5 seconds. The others were instructed to eat less than 50 millilitres in the first 30 seconds, and then to "continue at their own pace".

As you might expect, 27 per cent of the "Accelerated Eating" group of students reported ice-cream headaches, versus 13 per cent for the "Cautious Eating" group.

The Food Industry under the Microscope Part 3

The Low-GI Diet

You might have heard of the Glycemic Index (GI) and wondered what it was. In short, the GI measures what carbohydrates do to our blood sugar levels.

Professor Jennie Brand-Miller and her team from the University of Sydney have been researching the GI for years and are world leaders in the area. They've published many articles and books both for academics and general readers. A summary of the Low-GI Diet is another topic I wanted to highlight.

Carbohydrates 101

The word "carbohydrate" is pretty self-explanatory. "Carb" stands for carbon atoms, and "hydrate" means there is water. In other words, carbohydrates carry hydrogen and oxygen atoms in a 2:1 ratio, as in H_2O or water. And so carbohydrates are "hydrates" of carbon.

Chemists tell us there are four groups of carbohydrates: monosaccharides, disaccharides, oligosaccharides and polysaccharides.

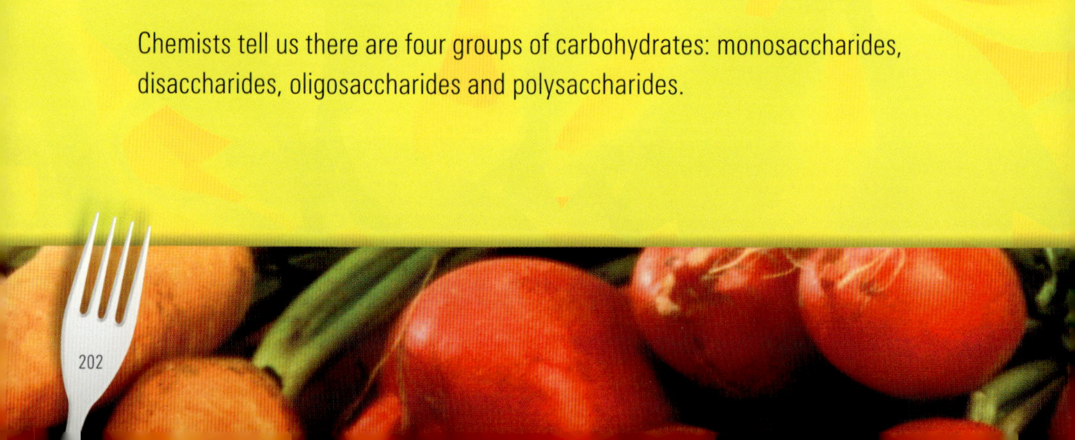

A monosaccharide, or simple sugar, is your basic carbohydrate. It's just a bunch of carbon, oxygen and hydrogen atoms arranged in a ring with five or six sides. In real life they are usually colourless, crystalline solids that dissolve in water. They include "glucose", "fructose", "galactose", "xylose", "ribose", and so on. You generally don't see naked monosaccharides outside a chemistry laboratory, but you get close with glucose jellybeans.

Disaccharides, on the other hand, are probably sitting on your kitchen table right now. The disaccharide "sucrose" is what you and I call "sugar" – the stuff you add to tea, coffee or porridge. Sucrose is made from glucose and fructose stuck together. The carbohydrate (or sugar) in milk, "lactose", is made from glucose and galactose stuck together. That's the pattern for disaccharides – two simple sugars stuck together.

What about oligosaccharides? Well, "*oligo*" means "few" in Greek, so an oligosaccharide has between 2 and 10 simple sugars stuck together. They can be found as "inulin" or "fructo-oligosaccharides" in onions, asparagus and artichokes. Oligosaccharides are also found on the membranes of cells in our body, such as the markers that give us the ABO blood groups.

Finally, we have the polysaccharides. They can have hundreds or thousands of simple sugars stuck together – sometimes in a single long chain, sometimes branching. Polysaccharides include starch (used to store energy), cellulose (used to make structures in plants) and chitin (used to make the shells of crabs).

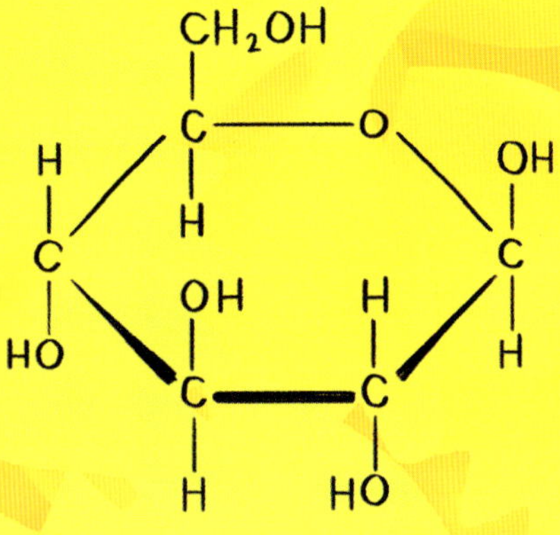

Glucose

GI 101

The word "glycemic" just means "related to glucose". The Glycemic Index tells you how quickly carbohydrates get broken down to simple sugars like glucose. Yep, carbohydrates get broken down into simple sugars at different speeds – and the Glycemic Index measures this.

Glucose (the simplest carbohydrate) is needed by all the cells in your body. Glucose goes from your food into your bloodstream. It is then pushed into your cells by insulin. Then it can be used for energy, stored as glycogen or converted into fat.

Now here's a very important point: high insulin levels stop a cell from using fat for energy needs. So it is important to avoid having high insulin levels all the time (which can happen by eating continually).

When a carbohydrate gets broken down quickly, your blood sugar level rises rapidly in a quick spike, insulin is pumped into your bloodstream to remove it, and so your blood sugar level quickly falls. Glucose has a rating of 100 on the Glycemic Index.

But a complex carbohydrate (for example, legumes such as soybeans or chickpeas) has a much lower GI. It breaks down slowly, so the simple sugars will be released slowly into your blood, as will the insulin. Your blood sugar level will rise gradually and then fall more slowly over a longer period of time – and it will not reach as high a peak as after eating a high-GI food. This is much better for you.

Glucose gets a GI rating of 100. By definition, a low-GI food is under 55, a moderate GI food is 56–69 and a high-GI food is 70 and above.

The Benefits of a Low-GI Diet

Carbs are found in foods from plant sources – like grains, cereals, vegetables and fruit – and they give you energy. The best eating plans *don't* cut out carbs. Instead, they advise good carb choices – plenty of complex low GI carbs.

A low-GI diet sits well alongside the eating advice in the Pollan and Kessler books because it encourages us to choose unprocessed foods – in other words, foods that slowly release energy.

If you can avoid eating highly refined carbohydrate meals and snacks, you'll also avoid cravings for foods that do not satisfy your hunger. This in turn will get you out of the loop of high sugar loads triggering peaks of insulin (which send blood glucose levels crashing, and leave you craving something else to eat).

Low GI = More Full Feeling

Sticking to a low-GI diet and exercising regularly also reduces your risk of getting Type 2 diabetes.

It's not only about the quantity of carbs – it's about the quality. Eating less food actually *lowers* your body's metabolic rate (BMR), which then *slows* your weight loss. But a low-GI diet lowers your BMR less than a low-fat diet does. This makes weight loss easier.

GI Specific Advice

A low-GI diet is consistent with current nutrition guidelines. Naturally, you still need to include lean protein in your diet.

Low-GI diets suit most people because you often don't have to change the structure of your regular meals. You just have to adjust the choices so they are more consistently low GI. For example, going low GI can be as easy as having pearl barley or noodles instead of white rice with your dinner.

Try to eat 2+ pieces of fruit per day and 5+ serves of vegetables. (Not potatoes, though, as they're often high GI.)

Eat about 5 serves of grains per day. Bread can be low GI, but white sliced bread is probably not. If you're unsure, check the GI values in printed tables or on food labels.

White rice tends to be high GI (but, surprisingly, sushi is low GI as the vinegar in the rice changes the GI index). Pasta tends to be low GI. Oats, wholegrain wheat, rye and all legumes are good low-GI choices.

Low-fat dairy scores lots of ticks on low-GI charts – generally you need 2–3 serves per day of dairy.

Go Nuts – but not too Nuts

Of course, you need to use common sense when using GI values to choose healthy foods. For example, potato chips are actually low GI! This is because the fat in which they are fried slows the breakdown of the carbohydrate in the potato. This doesn't make them a good choice for someone trying to lose weight – because the fat makes them much more energy dense.

Low-GI diets have been successfully developed for people with pre-diabetes, vegetarians, children and for anyone trying to lose weight. There are many, many publications giving detailed dietary advice depending on your needs and situation.

GI Jennie might LOL at some of the more crazy diets when hers is the last one standing.

CrO3
dil. H2SO4
> 90%
3-Menthene
OH
CrO3
CH3COOH
(-)-Menthol
PCl5
COOH
Menthyl chloride
Cl

Cool Mint

If you're ever in a field and come across Japanese mint or peppermint, pluck and then chew a few leaves and see what happens. Most likely, you'll feel a strange, cool sensation sweep across your mouth.

What's going on?

It was only in 2002 that scientists found out. Perhaps they'll be able to use this information to come up with new taste sensations, or even cure some diseases.

Menthol 101

There are about 7800 species in the Mint Order of Flowering Plants. Many types have given us essential oils which are used in beverages, in medicines, in perfumes, and to flavour various foods. The oils are usually found in little resin-laden dots in the leaves and stems.

The chemical which gives you that nice cooling effect in your mouth is called menthol (it's also known as peppermint camphor). For many centuries menthol was obtained from the oil of the Japanese mint plant, and it is now used in cosmetics, flavourings and cigarettes (to cool down the heat of the smoke dumping hot carcinogens into your system).

Menthol is a tiny molecule made up of only 31 atoms – ten carbon, 20 hydrogen and a single atom of oxygen ($CH_3C_6H_9(C_3H_7)OH$). In the same way that you can have left- and right- handed gloves, you can have left- and right- handed chemicals. It turns out that it's only the left-handed form of menthol that gives that mysterious cooling effect.

We get the sensation of "hot" and "cold" when special thermal receptors across our body are stimulated. These localised hot and cold spots on your skin were first reported back in 1882.

Slow Scientific Progress

By 1936 science had advanced to the stage that we could record electrical signals from a single thermosensitive nerve fibre in the tongue of a cat. By 1960 we could do similar research into the skin of humans.

We've still got a lot to learn about thermoreceptors, but it seems that on most of our body we have about seven times as many receptors to "cold" as we do to "hot". They seem to sit on skinny little nerves and their density varies. So while your lips might have 25 cold receptors per square centimetre, your middle finger might carry only 3 per square centimetre, and your chest might have fewer than 1. This pattern is also true of the warm receptors.

Greek Wisdom

Aristotle was Incredibly Insightful

When Aristotle wrote about the Five Senses, he described "sight" as the most supreme, as it gave the highest pleasure. He described "touch and sensing temperature" as the most basic sense, because it is essential for our survival.

All animals are sensitive to temperature, and insects especially so. Insects which suck blood have greater temperature sensitivity than the others. Indeed, lice and mosquitoes use the warmth of their victim's body as one of their ways to find more blood.

What is a "Receptor"?

Many of the nerves in the human body sense the outside world via receptors that work like a lock and key. The "lock" is usually a biggish protein that sits on the cell, and the "key" is the chemical that comes along and fits exactly into that (and no other) receptor. When the "key" goes into the "lock", the receptor starts firing and, very shortly afterwards, so does the nerve cell.

The receptor for both "menthol" and "the sensation of cold" is a protein made of 1104 amino acids, weighing around 128,000 daltons. In comparison, water weighs about 18 daltons, while insulin has a molecular weight around 6000 daltons.

In recent years we have discovered the receptor that senses both "cold" and "menthol". It's called TRPM8 (or Transient Receptor Potential Molecule Type 8). When this receptor senses cold temperatures between 8°C and 28°C, it fires off signals to your brain. And, sure enough, it does exactly the same thing when it senses the presence of menthol. It doesn't matter if you've got "cold" affecting this nerve, or "menthol" – either way, the TRPM8 receptor registers the sensation of "coolth". That's how menthol tricks your mind into feeling cold.

Peppermint Burp, not Fart

Peppermint does more than just give you a cool feeling in your mouth.

Peppermint releases the sphincter muscle separating the bottom of your oesophagus from the top of your stomach. This means that if you have a little peppermint after a big meal, instead of passing wind out through the back end you can surreptitiously burp it out through the top end.

Going Forward . . .

Perhaps all this research into menthol will help us treat a strange disease called Cold Allodynia. People afflicted with this disease have nerve damage and feel intense pain from something as mild as a cool breeze on their skin. We have already found that in some cases stimulating the menthol receptor can relieve the pain.

Thomas Hofmann and his colleagues from the German Research Centre for Food Chemistry are trying to separate the "coolth" of menthol from its "taste".

Every time you place peppermint on your tongue, you get both the sensation of "coolth" being sent to your brain as well as the refreshing taste of mint. Dr Hofmann has been fooling around with chemicals from roasted dark malt and has come up with 26 new chemicals that can give you the sensation of "coolth" – but without any of the taste. The best of these newly designed chemicals is 35 times better at cooling the mouth than menthol, and 250 times better at cooling the skin! Dr Hofmann believes that, within a few years, new non-minty cool sensations will start appearing in toothpastes, chocolates, citrus drinks, burn creams or – wait for it – drinking water.

Imagine it: you'll be able to drink water that your hands tell you is warm but your tongue tells you is cool. How's that for messing with your mind?

Ginger Beer Plant

These days, ginger beer is a sweet, ginger-tasting, non-alcoholic soft drink. It's made by adding sugar, ginger and carbon dioxide to water.

But in its shady past, ginger beer was very different. Back in the mid-1800s, not only could you make it in a jar on your kitchen shelf, it was also fizzy, alcoholic and cheap – a winning combination.

You started with a Ginger Beer Plant. This "plant" didn't have green leaves or a stem or a trunk – it was a slowly bubbling, gelatinous mess in a glass jar. People knew that if you fed your ginger beer plant with sugar, water and ginger, the mixture bubbled away happily, making alcoholic ginger beer.

There were two problems – nobody knew where the Ginger Beer Plant originally came from, and nobody knew how it worked.

In 1887 an English botanist named Harry Ward began many years of painstaking research. He never found out where the plant came from – in fact, we still don't know – but by 1892, he had solved the mystery of how it worked.

Two microorganisms were equally essential – a fungus (*Saccharomyces*) and a bacterium (*Lactobacillus*). Together, these microbes made carbon dioxide and alcohol – up to 11 per cent strength.

Other fermented drinks like kefir (a yoghurty drink) are also the result of a symbiosis between a yeast and a bacterium.

Sadly, none of Harry Ward's Ginger Beer Plants survive today.

Poo Brown, Wee Yellow

How come your food enters your body in so many different and attractive colours, but always comes out brown? And why does water enter your body clear, but always come out yellow?

The cause is the same for both brown poo and yellow wee – chemicals from dead red blood cells (RBCs). This story is fairly technical. But if you want to ignore the fine detail, here's the Take-home Message:

1 Poo is brown and urine is yellow because of a chemical that gets released when red blood cells are broken down.

2 This chemical goes through seven major "transformations" before it ends up as the brown chemical that gives the brown colour to faeces.

3 During the process of being "transformed", the chemical travels between the spleen, the bloodstream, the liver, the gall bladder and the duodenum before it finally exits your body into your toilet bowl.

4 During one of the intermediate stages, this chemical is yellow in colour. It "escapes" into the bloodstream but is picked up by the kidneys and ends up in your bladder, and again exits your body into the toilet bowl.

5 When making tea, do not boil the water for too long. In fact, people who grow tea (and love tea to pieces) just heat the water enough to almost (but not quite) boil it.

6 If you want to have a full and deep understanding, read on.

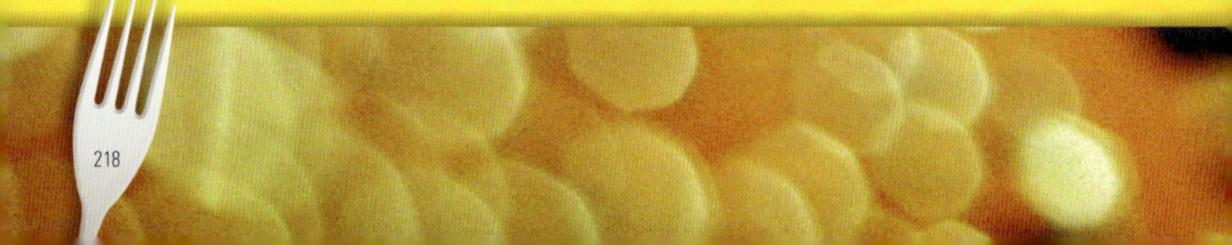

Red Blood Cells (RBCs)

This story begins in the bone marrow. It takes about seven days to make a red blood and the bone marrow expels about 2.4 million of these cells every second. RBCs take about 20 seconds to do a complete loop of the body. They get pumped out from the heart to the extremities, then flow back via veins into the heart and into the lungs to be oxygenated, and back into the heart again ready to be pumped out once more. Your blood is about 45 per cent cells (over 95 per cent of them RBCs) and 55 per cent salty water.

RBCs are much smaller than most of the other cells in the human body – about 7 microns across. (A micron is one millionth of a metre. For comparison, a human hair is about 70 microns across.) RBCs make up about one-quarter of the 100 trillion or so cells in your body – but they don't make up one-quarter of your weight because they are so small.

Each RBC carries about 270 million "haemoglobin" molecules. Between all of them they carry 2.5 grams of iron, which is about 65 per cent of your body's total iron stores.

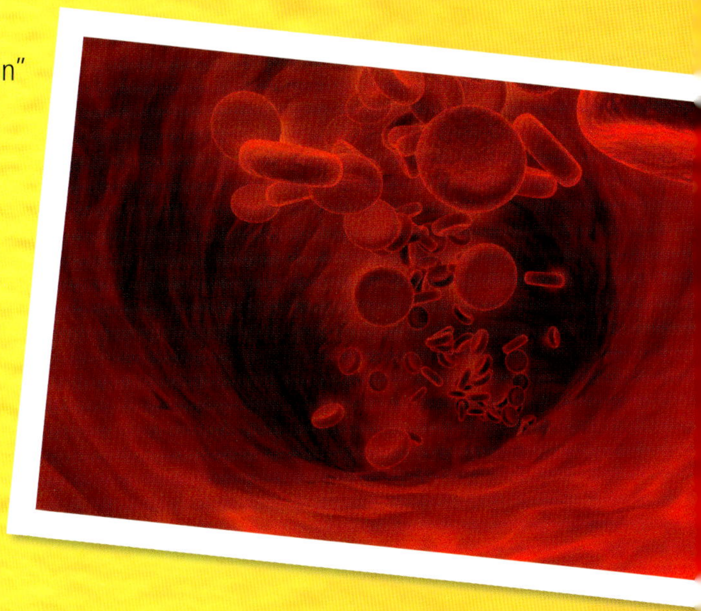

The Austrian molecular biologist Max Perutz worked out the molecular structure of haemoglobin in 1959. He received a Nobel Prize for this. Haemoglobin has two parts – "haem" and "globin".

I'll talk about the "globin" part first, just to get it out of the way. There are four "globin" molecules in each haemoglobin molecule. (Yep, they are kind of "globular".) Each "globin" is quite big, with a total molecular weight of about 68,000 daltons. They are made from various amino acids.

Now for the "haem" part, which ends up making poo brown and urine yellow. It's made of four rings that are joined together by short straight sections. Finally, the ends are joined together to make a loop, with the rings on the inside. Right in the middle is a single atom of iron. Haem is quite a small molecule – about 616 daltons.

The RBCs are very flexible. They have to be, because on each pass around the body, they have to squeeze through capillaries whose diameter is about 20 per cent smaller than the RBCs.

After about 100–120 days, the RBCs are getting old. For one thing, they get stiffer and find it hard to squeeze through the capillaries.

RBCs get recycled in the spleen.

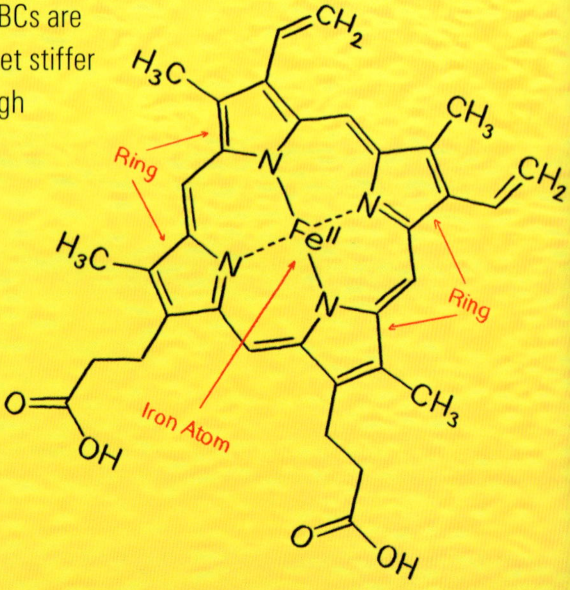

Tea and RBCs

Practically all animals with spines have some kind of red blood cell (with some version of haemoglobin inside). The only ones that don't are Crocodile Icefishes. They can do without red blood cells (and the haemoglobin) because they live in very cold water (and this relates to not boiling the water for too long when you make tea).

At normal human blood temperature, thanks to haemoglobin, the blood can carry 70 times more oxygen than can be carried by simply dissolving the oxygen in the blood. Oxygen dissolves really well in water at low temperatures (but really badly at high temperatures – and this also relates to tea). So at the very cold temperature that the Crocodile Icefishes live in, the salty water in the blood can carry enough oxygen to get by. The Crocodile Icefishes don't have to go to the huge metabolic expense of making red blood cells.

So what's the link to tea? Remember that oxygen dissolves really badly in water at high temperatures – so if you boil the water for a long time, you drive off lots of the oxygen. This means that there is very little oxygen left to react with the tea and bring out the delicate flavours.

So in tea-growing areas they conserve the oxygen by not even boiling the water. They get it almost (but not quite) to the boil – and then spend the extra time they save on savouring the delicate flavours of their lovely tea.

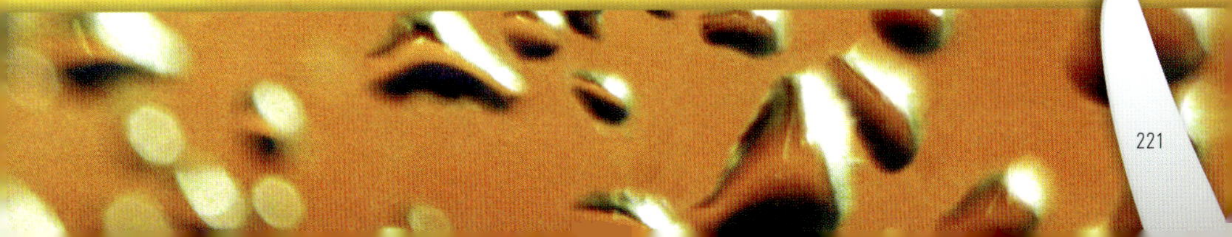

The Spleen

Have you every wondered exactly what your spleen does? It's an organ on the left side of your tummy, high up and under the ribs, about 11 cm long, and weighing between 150 and 200 grams. It does Immune System stuff, stores spare blood for use in an emergency – and it also "catches" the old and inflexible RBCs with mechanical filtration and breaks them down. It takes a lot of energy to make a haemoglobin molecule, so your body doesn't break it all the way down to its individual atoms.

The "globin" part of the molecule is broken down into individual amino acids. They are recycled to make proteins, including haemoglobin.

The "haem" part of the molecule is broken down into iron and "biliverdin" (that's the First Transformation). The "biliverdin" is turned to "bilirubin" (that's the Second Transformation). The shape of the bilirubin is basically the "haem" molecule unfolded into a straight line.

The bilirubin is released from the spleen into the bloodstream. But bilirubin doesn't dissolve well in water and, left to its own devices, tends to fall out of the blood and sit on the wall of blood vessels. When your spleen shoves bilirubin into blood, it finds and joins onto a common molecule called "albumin" (Third Transformation).

This combination (bilirubin attached to albumin) flows in the bloodstream to the liver.

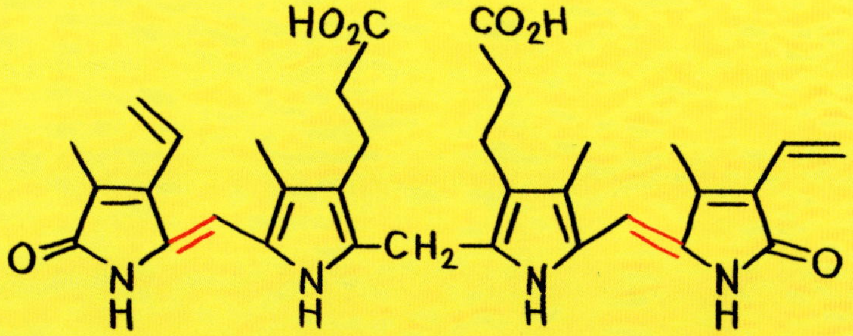

Albumen vs Albumin

These words are very similar, and both come from the Latin "albus", meaning "white". (In the Harry Potter movies Albus Dumbledore has both white hair and a white beard.)

"Albumen" is egg white or, more precisely, the sticky protein that is found in egg white.

"Albumin" is a more general term and refers to proteins that are soluble in water, and that will coagulate when heated. So "albumen" is one of the many types of "albumin".

Sorry about the confusion. The way that I remember the difference is by the fact that there is an "e" in both "egg" and "albumen".

The Liver

The liver grabs the combined molecule (bilirubin + albumin) from the bloodstream and removes the "albumin".

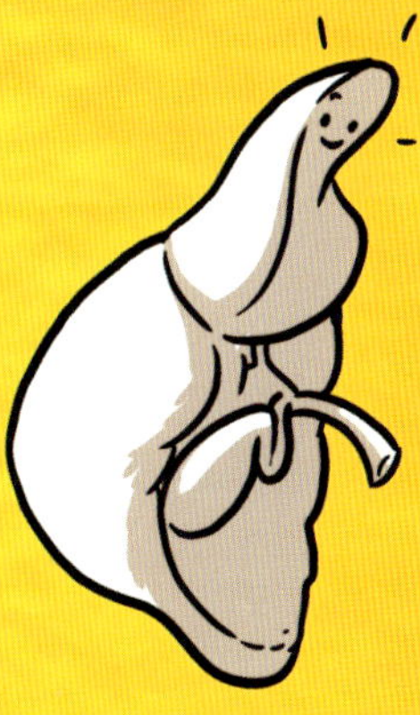

In the liver, two molecules of "glucuronic acid" are added to the bilirubin to make "bilirubin diglucuronide" (Fourth Transformation). This makes the bilrubin much more soluble in the blood. The liver then sends this chemical first to the gall bladder and then via bile ducts to the first part of your small intestine.

The Small Intestine

The bilbrubin diglucuronade enters the duodenum. One of its useful functions is that it helps dissolve fats that you have just eaten.

The bacteria in your gut turn it into yet another chemical, "urobilinogen" (Fifth Transformation). Most of this very useful chemical is recycled – the cells in the walls of your gut absorb it. It then goes back to the liver, back into the gall bladder and finally back into the gut. Once again, it will help dissolve fats that you have just eaten. During a typical meal the bile liquids will do this cycle two or three times.

What about the remaining 5 per cent that does not get reabsorbed? This urobilinogen gets turned into yet another chemical, "stercobilogen" (Sixth Transformation). Stercobilogen gets oxidised into "stercobilin" (Seventh Transformation) – and this is the magic chemical that gives your poo the brown colour.

Hurray! We have finally arrived at the end of our journey. It took a long time, but at least you now know the exact process that gives you brown poo.

Eat Poo = Yellow Face?

The Egyptian Vulture, an endangered fluffy white bird, is distinctly more attractive to its potential mate when it has a yellow face. The trouble is that the bird cannot make a yellow chemical with its own metabolism.

Enter the faeces of hoofed animals, such as cattle, deer and horses – the obvious solution. On one hand, these animal faeces carry some potential nutrients (such as the protein in the eggs and larvae laid by dung beetles and flies). And the faeces also contain lots of yellowish chemicals (luteins). On the other hand, the faeces are loaded with parasites that can potentially harm the vulture. Either way, the Egyptian Vulture takes its chances in its quest for yellow skin.

Maybe a yellow face on an Egyptian Vulture is a sign that the bird is so strong that it can eat animal poo and survive – and, therefore, will be a good potential mate. (What else would an Egyptian Vulture look for in its mate?)

Yellow Wee

Remember how the urobilinogen inside the gut leaves the gut wall and heads for the liver? Most of the urobilinogen gets captured by the liver, and is recycled into the gallbladder, and then back into the gut. But a small amount avoids this fate and enters the general blood circulation.

Somewhere along this pathway, urobilinogen gets transformed into a family of chemicals called "urobilins". Urobilins are yellow, and they make your urine yellow.

Your heart pumps out about 5 litres of blood every minute – and about 1 litre of that goes to the kidneys. The kidneys filter out the urobilins and send them to your bladder, where they then go to the toilet bowl.

And then, in most of Australia, the contents of the toilet bowl go to a beach near you.

Iron = Rust = Red Blood = Red Mars?

When iron rusts, it has a reddish colour.

So is blood "red" because the iron has combined with oxygen? Nope. Blood is red because the "haem" part of the haemoglobin molecule is "naturally" red. (I am taking the easy way out by saying "naturally" and not talking about "pi to pi* electronic transitions" – trust me, it's a lot easier this way.)

So what about the reddish colour of Mars? Yep, Mars is red for the same reason that the Australian outback is red – rust. In each case, iron has combined with oxygen to make rust, which is red.

Up Close and Personal

ANSWERS

One of the most important principles in explaining Science is:

"Anything, no matter how boring, looks better under a Scanning Electron Microscope (SEM)."

There are many SEM images artistically scattered throughout this book. They were taken for us by the clever people at the Australian Centre for Microscopy & Microanalysis, at the University of Sydney. We thank them enormously for their expertise, time and effort – in particular, Dr Ian Kaplin, Dr Errin Johnson, Dr Peter Hines, Dr Pat Trimby and Dr Jenny Whiting.

Colour My World . . . Why?

The wavelength of light is about half of a millionth of a metre – so we definitely can't use light to see anything smaller than that. Luckily, electrons have a much smaller wavelength – which is why we use Electron Microscopes to image really small stuff. (If you want to know more, check out Wikipedia, How Stuff Works, and so on.)

Unlike light, electron beams don't produce colour, so the SEM can only generate black and white images. The SEM images in this book have been colour manipulated and enhanced. For some, an attempt to represent the colours to resemble the 'item' being photographed has been made. At other times, well, let's just call it art . . .

The items have been magnifed up to hundreds of times – but the images labelled with ** have been magnified *thousands* of times. And here are the answers . . .

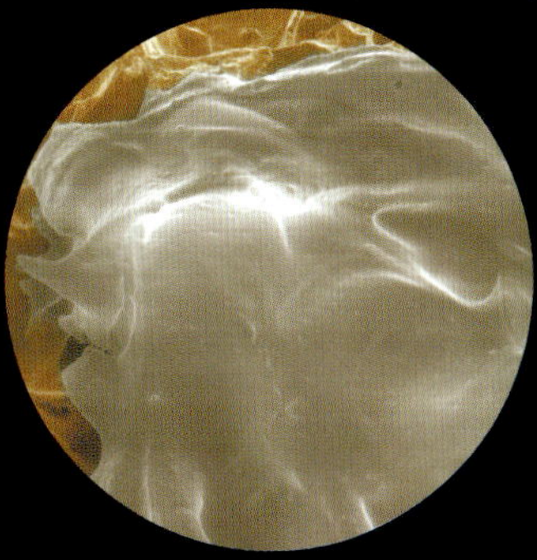

Page viii – Vegemite on wholemeal toast

Page 25 – Mould growing on orange peel

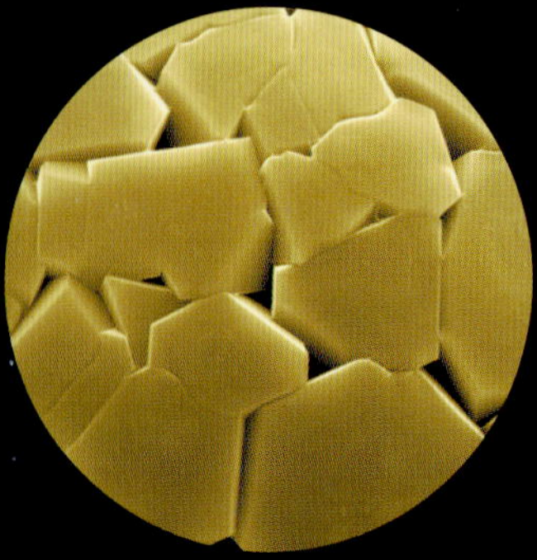

Page 27 – A single particle from a burning sparkler **

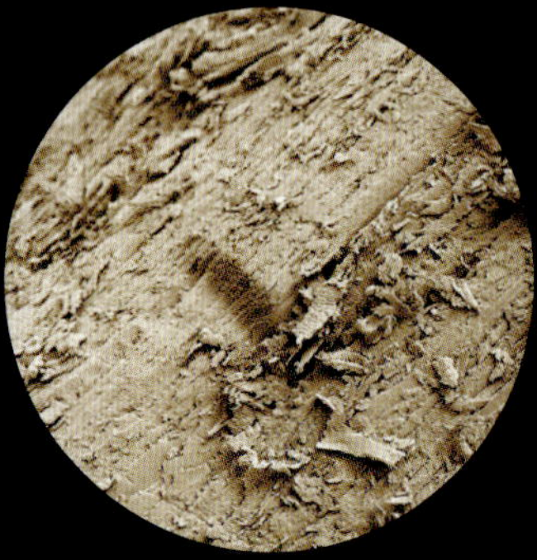

Page 42 – A toothpick

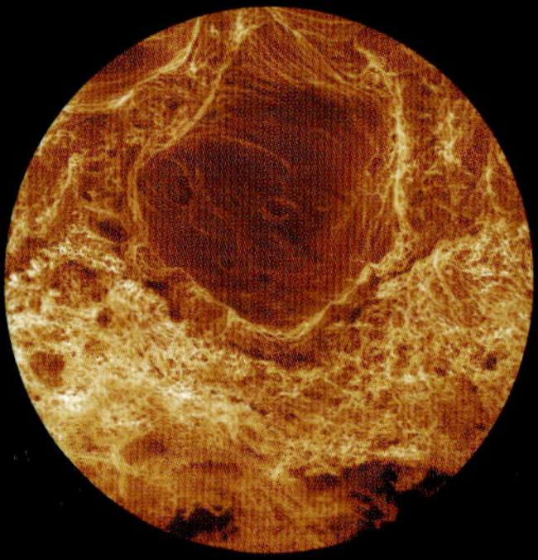

Page 51 – Mandarin peel

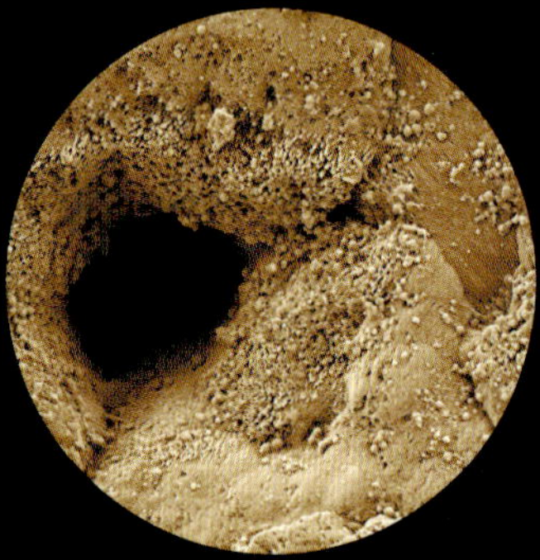

Page 61 – A pore in the surface of an eggshell **

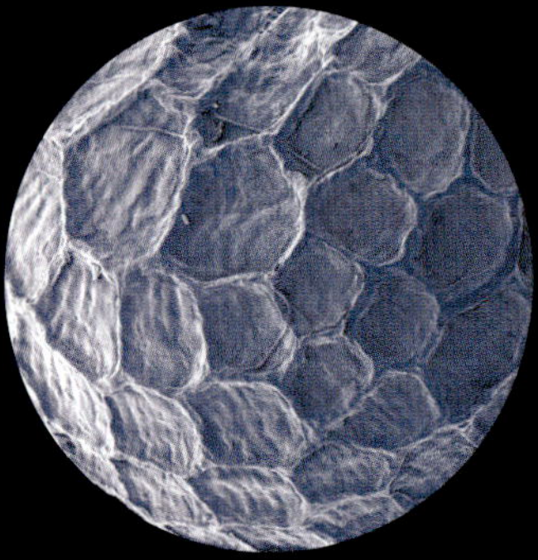

Page 94 – A poppy seed

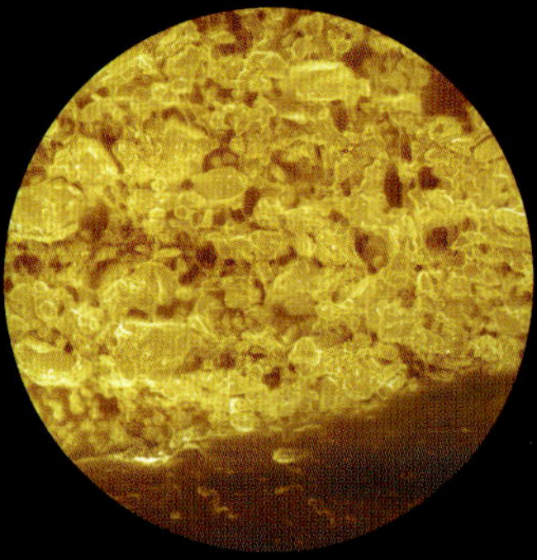

Page 109 – A licorice allsort

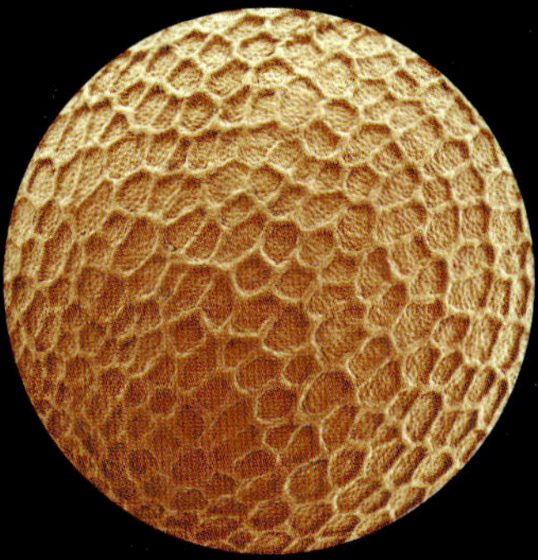

Page 117 – A mustard seed

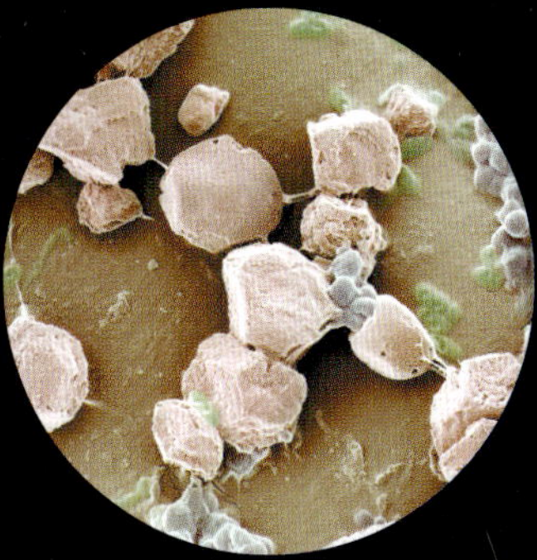

Page 129 – Kitchen sponge: sugar/ salt crystals (pink), bacteria and micro-organisms (blue) **

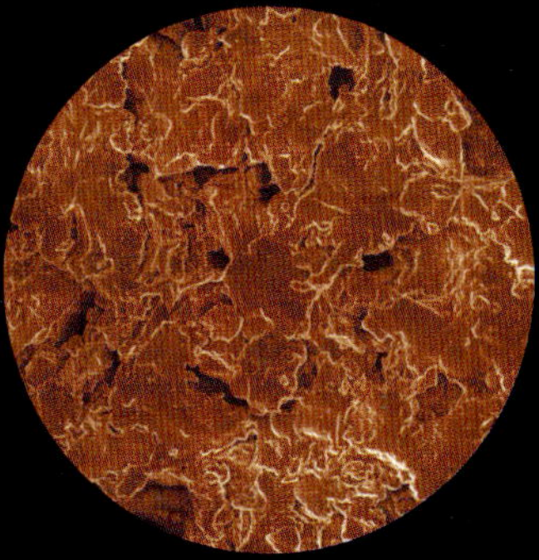

Page 137 – Cross-section of a Tim Tam

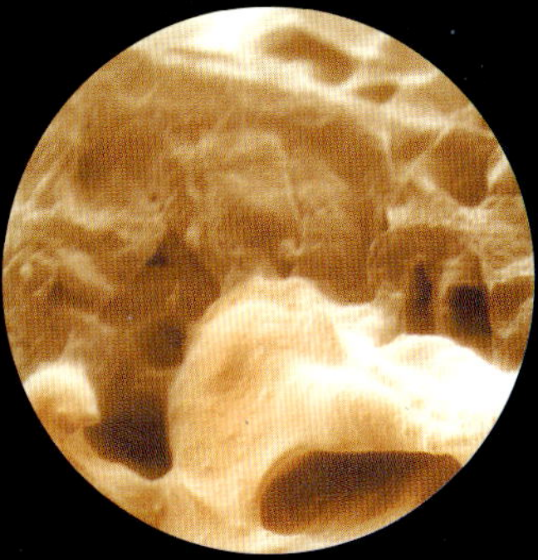

Page 156 – Peanut butter

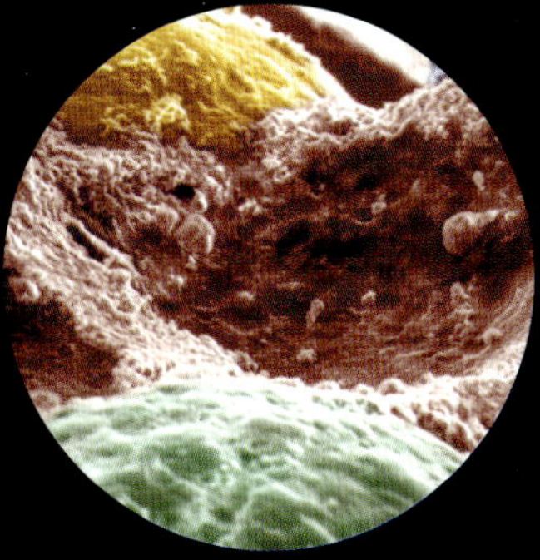

Page 167 – A chocolate freckle

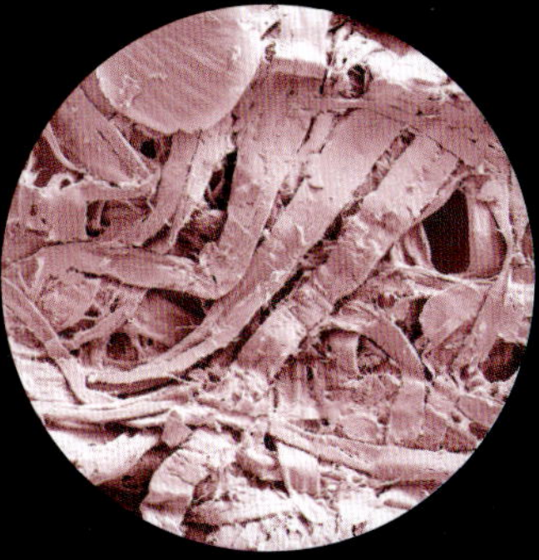

Page 187 – Toilet paper

Acknowledgments

In this Fine Book, I was very lucky to have some expert advice from Professor Jennie Brand-Miller (nutrition and Low GI, University of Sydney) and Dr Florent Angly (viruses, especially bacteriophages, University of Queensland). If there are any mistakes (and there always are), the Prof and Doc are like totally not to blame, dude.

Second, I would like to thank my Fabulous Colleague and Producer, Caroline "Bondi Mermaid" Pegram, who tore herself away from her beloved surf and multiple surfboards to whip this book into shape, with a goodly assortment of titles, suggestions, punchlines, more suggestions, other suggestions, different suggestions (you get the idea) . . .

Then, Big It Up for the Team at Pan Macmillan Australia: Rod "Rev Head" Morrison, Emma "Corros" Rafferty, Louise "What Happens on Tour . . ." Cornegé, Hayley "IT Girl" Crandell, Jane Hayes and Jessica Weir.

A big "Big It Up" for Douglas Holgate for once again enriching the look of this book with his fabulous illustrations.

The talented design crew at Xou Creative: Jon MacDonald, David Henley and Lucy Schuman. And my agent, Sophie "Country & Western" Hamley from The Cameron Creswell Agency.

I would also like to thank my Superb Skill in Procrastination. As a result, while the deadlines grew more menacingly closer, I found (nay, made) time to repair the cutlery drawer and then to rearrange all the cutlery, to clean the stove several times per day, and to load and unload our dishwashers continuously (even to the point that I would pull plates and cups out of the hands of my Family while they were still eating and drinking so that I could fill the hungry dishwashers). Read this and weep with sympathy for my Poor Publisher and Family. And, of course, I really did have to de-rust all the stainless-steel coathangers.

And finally, I must thank said Beloved Family, who changed their lives enormously so that I could write (a strange, solitary and consuming passion) in a lovely and calm environment – Lola, who is still our effervescent and unpredictable Baby; Little Karl, who is now quite Big; Alice, who is always quite perfect; and Mary (the most patient and beautiful wife, for whom I wished upon a star, and my dreams came true), who is the best sounding board and editor any author could hope for.

References

The Woman whose Life was Saved by a Poo Transplant

"How Microbes Defend and Define Us", by Carl Zimmer, *The New York Times*, 12 July 2010.

"*Clostridium difficile* Infection: An Overview of the Disease and its Pathogenesis, Epidemiology and Interventions", by V.K. Viswanathan et. al., *Gut Microbes*, July/August 2010, pages 234–242.

"Changes in the Composition of the Human Fecal Microbiome Following Bacteriotherapy for Recurrent *Clostridium difficile*–Associated Diarrhea", by Alexander Khoruts, MD, Johan Dicksved, PhD, Janet K. Jansson, PhD, and Michael J. Sadowsky, PhD, *Journal of Clinical Gastroenterology*, June 2010.

"The Human Gut Mobile Metagenome: A Metazoan Perspective", by Brian V. Jones, *Gut Microbes*, November/December 2010, pages 415–431.

"Same Poop, Different Gut", by Christina Luiggi, *The Scientist*, 3 November 2010.

Salmon Colour

"15 Colours of Salmon", by Elizabeth Cha, *Wired Magazine*, 24 February 2004.

The Tube of Life

Anatomy: A Regional Study of Human Structure, by Gardner, Grey and O'Rahilly, W.B. Saunders Company, USA, (4th edition), 1975.

Gray's Anatomy, by Roger Warwick & Peter L. Williams, Longman, UK, (35th edition), 1973.

Harrison's Principles of Internal Medicine, McGraw-Hill Inc., USA, (12th edition), 1991.

Human Physiology, by William F. Ganong, Appleton & Lange, USA, 1987.

Beer – Balm for Burns?

"Beer as a Burns Resuscitation Fluid", by Giles N. Cattermole et. al., *Emergency Medicine Australasia*, Vol. 22, 2010, pages 195–196.

The Fat Virus

"Fat Factors", by Robin Marantz Henig, *The New York Times*, 13 August 2006.

Shaken, not Stirred – Martinis are Forever

"Shaken, Not Stirred: Bioanalytical Study of the Antioxidant Activities of Martinis", by C.C. Trevithick et. al., *British Medical Journal*, 18–25 December 1999, pages 1600–1602.

"Two Parts Vodka, a Twist of Science", by Peter Meehan, *The New York Times*, 10 May 2006.

"Martini Mavens Go Totally Stir-Crazy", by Charles Perry, *The Los Angeles Times*, 26 December 2007.

"The Last Word: Stirring Stuff", *New Scientist*, 8 May 2010, page 57.

Cats Eschew Sweetness

"Pseudogenization of a Sweet-Receptor Gene Accounts for Cats' Indifference toward Sugar", by Xia Li et. al., PloS Genetics, Vol. 1(1) July 2005, pages 27–35.

Taste that Smell

The Odd Body: Weird and Wonderful Mysteries of Our Body Explained, by Dr Stephen Juan, HarperCollins Publishers, 1995, pages 9–12.

"A Nobel Smell" by Dr Robert Sylwester, November 2004, http://www.brain-connection.com/content/211_1

"Nobel Prizes", *Encyclopædia Britannica*, Encyclopædia Britannica 2006 Ultimate Reference Suite DVD.

"Learning to Smell the Roses: Experience-Dependent Neural Plasticity in Human Piriform and Orbitofrontal Cortices", by Wen Li et. al., *Neuron*, 21 December 2006, pages 1097–1108.

Antibiotics and the Gut

"Unraveling Gut Inflammation", by Warren Strober, *Science*, 25 August 2006, pages 1052–1054.

"Symbiotic Bacteria Direct Expression of an Intestinal Bactericidal Lectin", by Heather L. Cash et. al., *Science*, 25 August 2006, pages 1126–1130.

"Antibiotic Administration Alters the Community Structure of the Gastrointestinal Micobiota", by Courtney J. Robinson and Vincent B. Young, *Gut Microbes*, July/August 2010, pages 279–284.

"Antibiotics Play Hell with Gut Flora", *New Scientist*, 18 September 2010, page 17.

"Incomplete Recovery and Individualized Responses of the Human Distal Gut Microbiota to Repeated Antibiotic Perturbation", by Les Dethlefsen and David A. Relmana, *PNAS*, 15 March 2011, pages 4554–4561.

Popcorn and Pop Stars

"Popcorn Hazard", *New Scientist*, 30 April 1994.

"Popping the Question", by Julian South, David Hills and Ulrich Thimm, *New Scientist*, 16 September 1995.

"Hugely Corny", by Hazel Muir, *New Scientist*, 3 June 2000.

"Controlling the Size of Popcorn", by Daniel C. Hong and Joseph A. Both, *Physica A: Statistical Mechanics and Its Applications*, 15 January 2001, pages 3–4.

"Increasing the Size of a Piece of Popcorn", by Paul V. Quinn, Daniel C. Hong and Joseph A. Both, *Physica A: Statistical Mechanics and Its Applications*, 1 August 2005, pages 637–648.

"Doctor Links a Man's Illness to a Microwave Popcorn Habit", by Gardiner Harris, *The New York Times*, 5 September 2007.

"Popcorn", *Encyclopædia Britannica*, *Encyclopædia Britannica Ultimate Reference Suite*. Chicago: Encyclopædia Britannica, 2011.

The Strangers Within

"Host-bacterial Mutualism in the Human Intestine", by Fredrik Backhed et. al., *Science*, 25 March 2005, pages 1915–1920.

"Diversity of the Human Intestinal Microbial Flora", by Paul B. Eckburg, et. al., *Science*, 10 June 2005, pages 1635–1638.

"The Microbes Living Inside Us", by Laura Spinney, *New Scientist*, 15 August 2007.

"Slimming for Slackers", by Bijal Trivedi, *New Scientist*, 1 October 2005.

"Metagenomic Analysis of the Human Distal Gut Microbiome", by Steven R. Gill, et. al., *Science*, 2 June 2006, pages 1355–1359.

"The Next Human Genome Project: Our Microbes", by Emily Singer, *Technology Review*, 2 May 2007.

"The New Hygiene Hypothesis", by Emily Singer, *Technology Review*, 3 January 2008.

"Evolution of Mammals and their Gut Microbes", by Ruth E. Ley et. al., *Science*, 22 May 2008.

"The Inside Story", by Asher Mullard, *Nature*, 29 May 2008, pages 578–580.

"Fascinating Science, Compelling Cause", by V.K. Viswanathan, *Gut Microbes*, January/February 2010, page 3.

"A Human Gut Microbial Gene Catalogue Established by Metagenomic Sequencing", by Junjie Qin et. al., *Nature*, 4 March 2010, pages 59–64.

"Genetic Pot Luck", by Justin L. Sonnenburg, *Nature*, 8 April 2010, pages 838–838.

"Transfer of Carbohydrate-Active Enzymes from Marine Bacteria to Japanese Gut Microbiota", by Jan-Hendrik Hehemann, *Nature*, 8 April 2010, pages 908–912.

"Community Health Care: Therapeutic Opportunities in the Human Microbiome", by Justin L. Sonnenburg and Michael A. Fischbach, *Science Translational Medicine*, 13 April 2011, 78ps12, pages 1–6.

"The Human Gut Mobile Metagenome: a Metazoan Perspective", by Brian V. Jones, *Gut Microbes*, November–December 2010, pages 415–431.

"Bacterial Ecosystems Divide People into 3 Groups, Scientists Say", by Carl Zimmer, *The New York Times*, 20 April 2011.

"Enterotypes of the Human Gut Microbiome", by Manimozhiyan Arumugam et. al., *Nature*, 20 April 2011, pages 174–180.

The Perfect Cheese Sandwich

"How to Cut the *Perfect Cheese Sandwich*", *New Scientist* (Feedback), 8 November 2003.

Fat Germs

"The Gut Microbiota as an Environmental Factor that Regulates Fat Storage", by Fredrik Backhed et. al., *Proceedings of the National Academy of Sciences*, 2 November 2004, pages 15718–15723.

"Host-Bacterial Mutualism in the Human Intestine", by Fredrik Backhed et. al., *Science*, 25 March 2005, pages 1915–1920.

"Obesity Alters Gut Microbial Ecology", Ruth E. Ley et. al., *Proceedings of the National Academy of Sciences*, 2 August 2005, pages 11070–11075.

"Slimming for Slackers", by Bijal Trivedi, *New Scientist*, 1 October 2005.

"Fat Factors", by Robin Marantz Henig, *The New York Times*, 13 August 2006.

"Obesity and Gut Flora", by Matej Bajzer and Randy J. Seeley, *Nature*, 21/28 December 2006, pages 1009, 1010.

"Human Gut Microbes Associated with Obesity", by Ruth E. Ley et. al., *Nature*, 21/28 December 2006, pages 1022–1023.

"An Obesity-Associated Gut Microbiome with Increased Capacity for Energy Harvest", by Peter J. Turnbaugh et. al., Nature, 21/28 December 2006, pages 1027–1031.

"Is Obesity an Oral Bacterial Disease?", by J.M. Goodson et. al., *Journal of Dental Research*, July 2009, pages 519–523.

"The Effect of Diet on the Human Gut Microbiome: A Metagenomic Analysis in Humanized Gnotobiotic Mice", by Peter J. Turnbaugh et. al., *Science Translational Medicine*, 11 November 2009.

"Gut Check: Testing a Role for the Intestinal Microbiome in Human Obesity", by Jeffrey S. Flier et. al., *Science Translational Medicine*, 11 November 2009.

"Same Poop, Different Gut", by Christina Luiggi, *The Scientist*, 3 November 2010.

Marshmallow Moments

"Marshmallows and Public Policy", by David Brooks, *The New York Times*, 7 May 2006.

Panda's Puzzling Palate

"Evolution of Mammals and their Gut Microbes", by Ruth E. Ley et. al., *Science*, 22 May 2008.

"What is Black and White and a Puzzle all Over?", by V.K. Viswanathan, *Gut Microbes*, May/June 2010, page 3.

Saccharin Rots Rats' Regulation

"A Role for Sweet Taste: Calorie Predictive Relations in Energy Regulation by Rats", Susan E. Swithers, PhD and Terry L. Davidson, PhD, Purdue University; *Behavioral Neuroscience*, Vol. 122, No. 1, February 2008.

The Food Industry under the Microscope Part 1

In Defence of Food

In Defense of Food: An Eater's Manifesto, by Michael Pollan, Penguin Books Ltd, USA, 2008.

Peppery Wine Perfume

"From Wine to Pepper: Rotundone, an Obscure Sesquiterpene, is a Potent Spicy Aroma Compound", by Claudia Wood et. al., *Journal of Agriculture and Food Chemistry* 2008, Vol. 56, 3738–3744.

Bacteria Burger

"Pathogens in our Pork", by Nicholas D. Kristof, *The New York Times*, 15 March 2009.

"Cows on Drugs", by Donald Kennedy, *The New York Times*, 18 April 2010.

"Multidrug-Resistant Staphylococcus aureus in US Meat and Poultry", by Andrew E. Waters et. al., *Clinical Infectious Diseases*, 15 May 2011, pages 1227–1230.

Coffee Good Enough *Not* to Drink

"Is a Sniff of Coffee as Good as a Sip?", *New Scientist*, 14 June 2008, page 16.

Onions and Tears

"Crying Behavior and Psychiatric Disorder in Adults. A Review", by Vikram Patel, *Comprehensive Psychiatry*, 1993, Volume 34, pages 206–211.

"Tear-Inducing Onions Get the Chop", by James Randerson, *New Scientist*, 16 October 2002.

"An Onion Enzyme that Makes the Eyes Water", by S. Imai et. al., *Nature*, 17 October 2002.

"Differential Sex-Independent Amygdala Response to Infant Crying and Laughing in Parents Versus Non-Parents", by Erich Seifritz, et. al., 2003, *Biological Psychiatry*, Volume 54, 2003, pages 1367–1375.

"Why Do We Cry", by Chip Walter, *Scientific American Mind*, December 2006, pages 44–51.

"Crying as a Multifaceted Health Psychology Conceptualisation: Crying as Coping, Risk Factor, and Symptom", by Ad Vingerhoets and Lauren Bylsma, *The European Health Psychologist*, December 2007, pages 68–74.

"When is Crying Cathartic", by Lauren M. Bylsma et. al., *Journal of Social and Clinical Psychology*, 2008, Vol. 27, No. 10, pages 1165–1187.

"Is Crying Beneficial?", by Jonathan Rottenberg, Lauren M. Bylsma, and Ad J. J. M. Vingerhoets, *Current Directions in Psychological Science*, 2008, Vol. 17, No. 6, pages 400–404.

"Slow Cook Onions, and the Results Are Delicious", by Russ Parsons, *The Los Angeles Times*, 21 January 2009.

"Onion", *Encyclopædia Britannica*, *Encyclopædia Britannica*, *Ultimate Reference Suite*. Chicago: Encyclopædia Britannica, 2011.

"Introducing the Things of Spring", by Gail Collins, *The New York Times*, 29 April 2011.

"The Grand Delusion", by Graham Lawton, *New Scientist*, 14 May 2011, pages 35–41.

Crunching Chips

"Testing Whether the Crunch is All it's Cracked Up to Be", by Harold McGhee, *The New York Times*, 4 July 2007.

Teaspoons Vanish

"The Case of the Disappearing Teaspoons: Longitudinal Cohort Study of the Displacement of Teaspoons in an Australian Research Institute", by Margaret E. Hellard et. al., *British Medical Journal*, Vol. 331, 24–31 December 2005, pages 1498–1500.

The Food Industry under the Microscope Part 2

The End of Overeating

The End of Overeating: Controlling the Insatiable American Appetite, by David A. Kessler, Rodale Inc, USA, 2009.

Chocolate Not Cheery

"Bitter Shock for Chocaholics", by Kathryn Eccles, *Medical Observer*, 14 February 2006, page 12.

The Smell of Asparagus in the Morning

"On the Analysis of the Blood and Urine in Health and Disease. With Directions for the Analysis of Urinary Calculi", by G.O. Rees, *The Lancet*, 1836, i: pages 806–807.

"Food Idiosyncrasies: Beetroot and Asparagus", by S.C. Mitchell, *Drug Metabolism And Deposition*, Vol. 29, Number four, Part two, 2001, pages 539–543.

"Make it Snappy, but Not Too Snappy", by Marian Burros, *The New York Times*, 5 June 2007.

"Excretion and Perception of a Characteristic Odor in Urine After Asparagus Ingestion: A Psychophysical and Genetic Study", by Marcia Levin Pelchat et. al., *Chemical Senses*, 36:9–17, 2011.

The Kilogram is Losing Weight

"The Kilogram isn't What it Used to Be—It's Lighter", by Dava Sobel, *Discover*, March 2009.

Vodka has No Calories?

What Einstein Told his Cook, by Robert L. Wolke, W. W. Norton & Co, New York, 2002, pages 244–245.

"Carbohydrate Trends in Alcoholic Beverages", by Damon Brown, *Journal of the American Dietetic Association*, June 2005, pages 880–881.

"Alcoholic Beverage", *Encyclopædia Britannica*, Encyclopædia Britannica 2006 Ultimate Reference Suite DVD.

"Vodka", *Encyclopædia Britannica*, Encyclopædia Britannica 2006 Ultimate Reference Suite DVD.

"Striking Back at Soft Drinks, Bacardi Plans Low-Cal Rum", by Melanie Warner, *The New York Times*, 26 January 2006.

Ice-cream Headache

"Lick and Learn", by Justin Mullins, *New Scientist*, 26 July 1997, pages 40–41.

"Ice Cream Headache", by J.W. Sleigh, *British Medical Journal*, 6 September 1997, page 60.

"Ice Cream Evoked Headaches (ICE-H) Study: Randomised Trial of Accelerated Versus Cautious Ice-cream Eating Regimen", by Maya Kaczorowski and Janusz Kaczorowski, *British Medical Journal*, 21–28 December 2002, pages 1445–1446.

"Ice-cream Headache – A Large Survey of 8359 Adolescents", by Jong-Ling Fuh et. al., *Cephalalgia*, December 2003, pages 977–981.

The Food Industry under the Microscope Part 3

The Low-GI Diet

The Low GI Diet: 12-week action plan, by Jennie Brand-Miller et. al., Hachette Australia, 2004.

Cool Mint

"Systematic Studies on Structure and Physiological Activity of Cyclic r-Keto Enamines, a Novel Class of 'Cooling' Compounds", by Harald Ottinger, et. al., *Journal of Agricultural and Food Chemistry*, 6 November 2001, pages 5383–5390.

"Identification of a Cold Receptor Reveals a General Role for TRP Channels in Thermosensation", by David D. McKemy et. al, *Nature*, 10 February 2002, pages 52–58.

"Cold Comfort", *Nature*, 31 August 2006, page 960.

"The Menthol Receptor TRPM8 is the Principal Detector of Environmental Cold", by Diana M. Bautista et. al., *Nature*, 12 July 2007, pages 204–208.

"Channeling Cold Reception", by Bernd Nilius et. al., *Nature*, 12 July 2007, pages 147–148.

Ginger Beer Plant

"Marriage of Equals", by Gail Vines, *New Scientist*, 28 September 2002.

Poo Brown, Wee Yellow

"An Unusual Source of Essential Carotenoids", by J.J. Negro et. al., *Nature*, 25 April 2002, pages 907–908.

ALSO TRY!

Dr Karl's

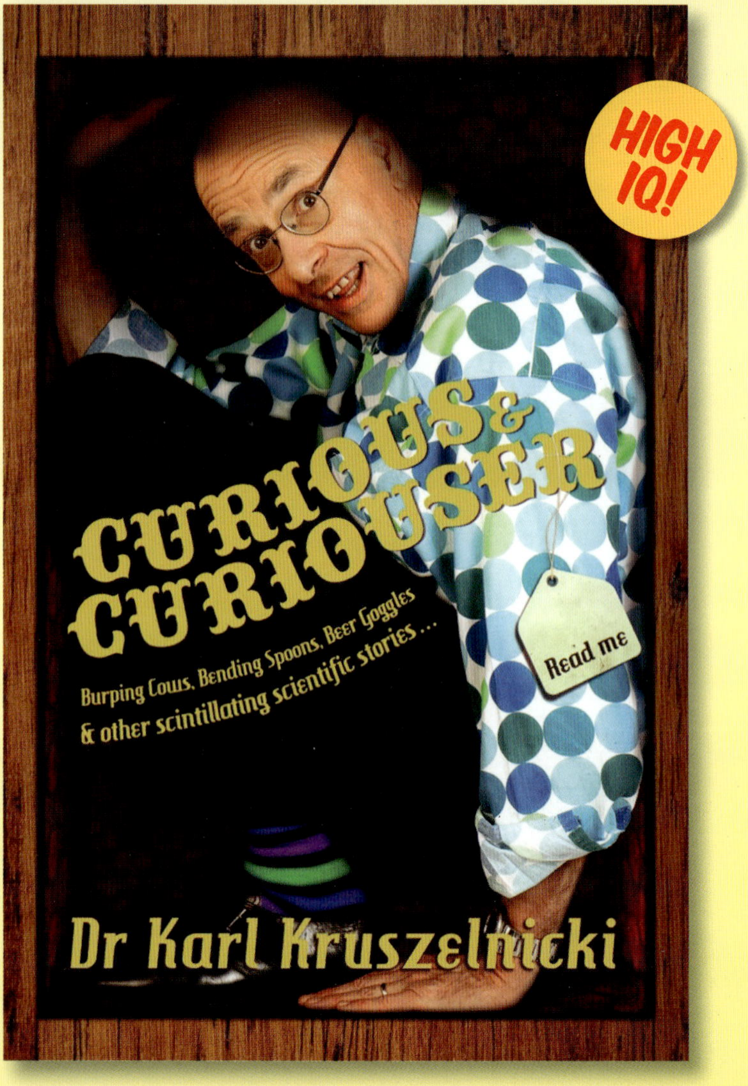

ISBN: 9781405040020

AND FOR THE KIDS . . .

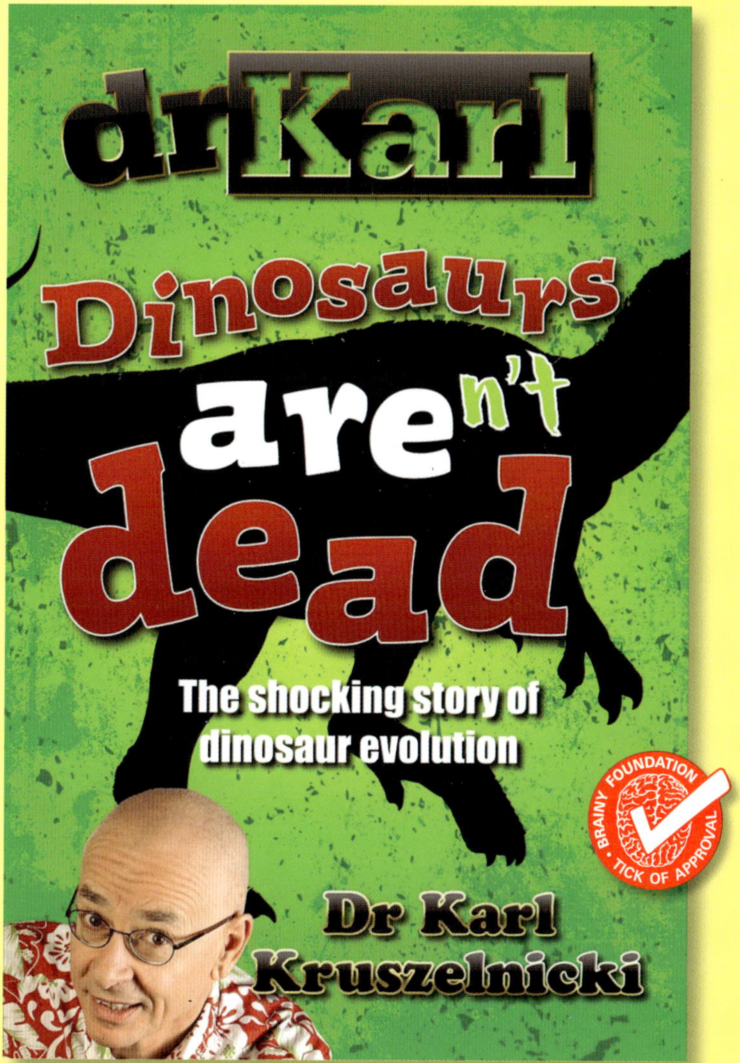

ISBN: 9780330425797

VISIT DRKARL.COM

Discover a little more about Dr Karl's world.

- Read his full bio in "About Karl"
- Check out the books/DVDs in "The Karl Collection"
- Take a look at Dr Karl's holiday snaps in "Picture This"
- See what "Things Karl Likes"
- Subscribe to the blogcasts and podcasts

✓ ADDED BRAIN ENHANCERS

✓ BREAKFAST OF BOFFINS!

✓ HIGH FALLUTIN'

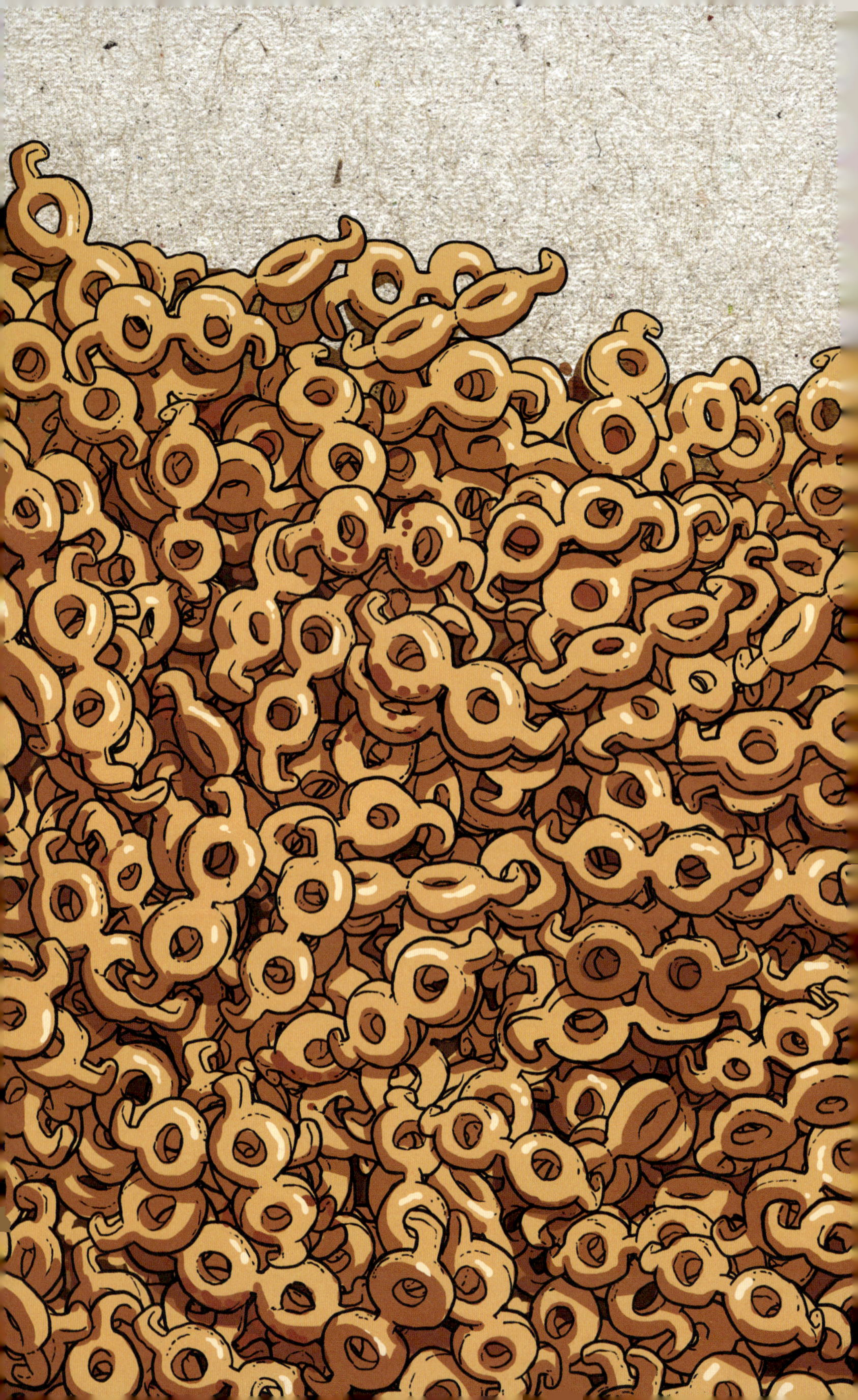